THE HERBARIUM HANDBOOK

THE HERBARIUM HANDBOOK

Edited by

Diane Bridson and **Leonard Forman**

Revised Edition

ROYAL BOTANIC GARDENS
KEW

© Copyright The Board of Trustees of The Royal Botanic Gardens, Kew
1989, 1991, 1992

First Published 1989

Reprinted with minor corrections 1991

Revised edition 1992

Typeset at the Royal Botanic Gardens, Kew by Pam Arnold, Christine Beard, Margaret Newman, Pam Rosen and Helen Ward.

Printed in Great Britain by
Whitstable Litho Printers Ltd.

ISBN 0 947643 45 1

THE HERBARIUM HANDBOOK

Edited by

Diane Bridson and **Leonard Forman**

Revised Edition

ROYAL BOTANIC GARDENS
KEW

© Copyright The Board of Trustees of The Royal Botanic Gardens, Kew
1989, 1991, 1992

First Published 1989

Reprinted with minor corrections 1991

Revised edition 1992

Typeset at the Royal Botanic Gardens, Kew by Pam Arnold, Christine Beard, Margaret Newman, Pam Rosen and Helen Ward.

Printed in Great Britain by
Whitstable Litho Printers Ltd.

ISBN 0 947643 45 1

FOREWORD TO FIRST EDITION

The Royal Botanic Gardens, Kew frequently receives requests to give training in herbarium curation and management to herbarium workers from many countries. There is therefore an obvious world-wide need for such training and as a result an 'International Diploma Course in Herbarium Techniques' was organized and held at Kew in 1987 and again in 1988. The first two courses were extremely successful, with participants coming from many parts of the world. It is planned to continue holding annual courses, but space is limited and so far the number of applicants greatly exceeds that which can be accepted in any one year. In view of this interest, and the extensive preparations that were made, it was felt that the main information imparted during the course should be published both for instruction and for reference. Many staff members of the Kew Herbarium and other departments of the Royal Botanic Gardens were involved in giving lectures and demonstrations and in producing lecture-notes and other explanatory pamphlets. The present handbook has been based largely on the documents provided during the Kew course. In many cases the original format and wording was not appropriate for the purpose of this handbook, and so revision, sometimes with extensive re-writing, has been necessary. Furthermore, since the original texts were written by many different authors, there were differences in style and presentation. An attempt has therefore been made to bring some consistency of style to the present text.

This handbook deals primarily with the technical aspects of herbarium work: the preparation, housing, preservation and organization of herbarium collections, and associated subjects. The science of taxonomy itself, i.e. how to carry out taxonomic research, is beyond the scope of this handbook, but many aspects of taxonomy including the terminology used in nomenclature and typification must be understood by herbarium technicians, and therefore a general background to these topics is given. Herbarium technicians generally learn the taxonomy they need to know, or want to know, while working under the supervision of trained botanists. Users of this handbook who need an introduction to the subject are recommended to consult Jeffrey (1982). Experienced technicians with a good 'eye' for plants and a real interest in them can gradually build up a knowledge of plant families and genera, and will eventually be able to recognize many of them. Plant identification and knowing what 'spot' characters to look for when sorting into families can only be learnt satisfactorily by practice under the guidance of an experienced botanist. Practical sessions on plant identification were held during the Kew course, but since practical training is essential, the subject is not dealt with in the present work. The procedures and materials described or recommended are inevitably influenced to some extent by those in use at Kew. But the recommendations can be adapted as necessary, and alternatives are mentioned. In a major herbarium such as Kew with world-wide collections, some of the procedures are necessarily rather complicated, but these can be simplified as appropriate in smaller institutions. Different traditions and habits develop in different institutions, and even within one institution, so there will be no unanimous acceptance of every procedure described. But for smaller herbaria needing guidance it is hoped that this handbook compiled by experienced herbarium staff will prove useful.

The editors would be grateful to receive any comments or suggestions for the improvement of future editions.

Acknowledgements

The writing of this handbook has involved the cooperation of many people. Thanks are due to the following members of the Kew staff for their contributions to the text:- Barry Blewett, Peter Boyce, Diane Bridson, Derek Clayton, Tom Cope, Jill Cowley, Peter Edwards, David Field, John Flanagan, Carol Furness, David Goyder, Christopher Grey-Wilson, Nigel Hepper, Vicki Humphrey, Charles Jeffrey, Mike Lock, Mike Maunder, Simon Mayo, Barbara Parris, David Pegler, Martin Sands, David Simpson, Roger Smith, Brian Spooner, Brian Stannard, Milan Svanderlik and Jeffrey Wood. Other colleagues have made useful suggestions and their help is much appreciated.

Acknowledgement is due to the Department of Cryptogams, British Museum (Natural History) who kindly provided chapter 31 on the collection and curation of bryophytes. The Department would always be grateful to receive any collections of bryophytes.

Thanks are due to Sally Dawson for the major part she played in the preparation of the illustrations, to Sarah Oldridge for figs. 49–52 and to Media Resources (I.E.D.) for figs. 4, 14 & 26.

I should like to give my special thanks to Martin Sands and Barrie Blewett who bore the main burden of organizing the first course, finding and solving the snags as they went along; to Diane Bridson and Barrie Blewett who developed the improved second course; and finally to Leonard Forman and Diane Bridson for all their efforts in preparing this publication — truly a team effort.

G.Ll. Lucas
Keeper of the Herbarium and Library
Royal Botanic Gardens, Kew

FOREWORD TO THE REVISED EDITION

The UN Biodiversity Convention proposed at the Earth Summit in Rio de Janeiro in June 1992 highlights the growing and widespread interest in biodiversity. The vital rôle that herbaria play in underpinning such projects is also becoming increasingly appreciated by a wider audience. The growing importance of herbaria and the information they contain cannot be overestimated. In order to meet future needs, therefore, it is essential that adequate human and financial resources are provided for their development and maintenance.

Three aspects of herbarium management deserve special attention:

1) The permanent preservation and management of collections and how this can best be achieved at realistic costs.

2) Adding to the collections. It is important that herbaria are enriched with specimens of high quality and not filled with poorly collected and inadequately documented material.

3) Understanding how to extract, use and make widely available the data associated with the specimens. International databases are now becoming a reality, and it is therefore important that those involved with data capture are able to understand fully the significance of specimen labels and the information they contain.

The Herbarium Handbook pays special attention to these three aspects, particularly the third, as so little has been published to guide the technician or junior taxonomist in this essential matter.

Since the publication of the first edition, International Diploma Courses in Herbarium Techniques have been held at Kew in 1990, 1991 and 1992. The experience gained from these courses and the comments and opinions expressed by students, reviewers and colleagues have been invaluable in producing this revised edition.

I especially wish to thank the Keeper, Stephen Blackmore, and his colleagues in the Department of Botany, Natural History Museum, London for their kind support and assistance. In particular, Jenny Moore wrote the chapter on the larger algae, and Alan Harrington contributed to the bryophyte chapter. This is one of those pleasing examples of how our two institutes work closely together for botany, and taxonomy in particular.

Finally I want to thank and recognize the dedicated attention to detail of Mrs Diane Bridson and Mr Leonard Forman in drafting and checking all the additions and changes for this edition. The quality of the product speaks for itself.

G.Ll. Lucas
Keeper of the Herbarium and Library
Royal Botanic Gardens, Kew

INTRODUCTION TO REVISED EDITION

The first edition of the Herbarium Handbook had a very favourable reception and the steady demand for copies soon led to a reprint, stocks of which are now exhausted. This has encouraged us to prepare a revised and considerably enlarged edition. There are two entirely new chapters: one on the conservation of herbarium sheets, including their cleaning and repair, and the division of sheets bearing mixed collections; the other on the collection and curation of the larger algae. The chapter on computers has been much expanded in response to requests for greater coverage of this increasingly important aspect of herbarium work. This chapter, it should be noted, is written for those with little or no knowledge of computers rather than for the computer-literate. It provides background information and general advice to help those planning to set up a new system for herbarium use. New regulations in the U.K. concerning the use of substances hazardous to health have caused considerable alterations to the sections on spirit collections and collecting into liquid preservatives. A distinction is now made between 'fixatives' and 'preservatives'. Cautionary advice has also been given for the use of pesticides, and the chapter on 'Pests and treatments' has been greatly expanded. Other chapters have been revised and new information added. The chapter on 'Collecting and preserving specimens' has further data on arranging and preparing the specimens, and now includes an account of the 'Polish press method' of drying. The coverage on fungi has been revised and enlarged and includes new illustrations demonstrating the principal types of Basidiomycetes and Ascomycetes. The improved text on economic botany now gives a classification of the different kinds of plant uses, collecting methods and instructions on recording data appropriate to the type of use.

'Collectors, itineraries, maps and gazetteers' is now more informative and includes useful hints as well as an enlarged section of references. Throughout the book bibliographical information has been increased and brought up-to-date. It should be noted that the references at the end of each chapter include 'further reading' as well as literature referred to in the text. Users of the Handbook who wish to learn more about taxonomy are recommended to consult De Vogel (1987) and Stace (1989) in addition to Jeffrey (1982).

It has not proved practical to include a list of firms specializing in archivally approved materials. If difficulty is experienced in obtaining or choosing products of conservation quality, contact should be made with the local conservators' or curators' group or society. Such organizations are often based in National Museums (of both natural history and the arts) and may well be affiliated to international societies such as ICOM–CC (International Council of Museums – Committee for Conservation), SPNCH (Society for the Preservation of Natural History Collections) or CIN (Conservation Information Network), which maintains an on-line Suppliers Database available to subscribers.

Acknowledgements

We are greatly indebted to our colleagues at Kew for their help in the preparation of this new edition. The writing of the computer chapter proved to be a major task; we thank Nicholas Hind for his preliminary notes, Bill Loader and his team in the Computer Section, Bob Allkin, Don Kirkup and Dave Simpson for reading the manuscript and contributing invaluable comments, and Bob Allkin for providing the references.

We are grateful to David Pegler, Brian Spooner and Thomas Laessøe of the Mycology Section for their considerable revision of the text relating to fungi (chapters 17 and 35). Special thanks are due to Francis Cook for the substantial body of new data contributed on economic botany (39). We are indebted to Charles Jeffrey for making available his guidance notes on the practice and ethics of separating mixed sheets (10) and to Mike Lock for the additional data on soil types etc. (40).

We wish to thank in particular the following for additions, amendments and suggestions to other chapters: from the Herbarium – Sally Bidgood (10 & 24), Peter Boyce (17), Phillip Cribb (13 & 31), John Dransfield (17), Carol Furness (21 & 23), Sue Holmes (30 & 36) and Nigel Taylor (30); from the Library – Leonora Thompson (22) and from the Library Preservation Unit – Kate Edmondson (5 & 14); colleagues from the Jodrell Laboratory (21 & 23) with special thanks to Tony Cox on information on material for DNA studies, and to Simon Linington and Roger Smith (37); from the Living Collections Department – Mike Maunder (36) and from the Conservation Section – Noel McGough (41). Thanks are due to Media Resources for fig. 21 and plate 1, and to Sally Dawson for figs. 26, 46–7, 49, 54–59, 61–63.

We are most grateful to reviewers and to those students from the 1990 and 1991 courses who have provided constructive criticism; we have acted on their suggestions where possible. We are especially indebted to Jenny Moore of the Department of Botany, Natural History Museum, for contributing the new chapter on larger algae (34) and to Alan Harrington for corrections to the chapter on bryophytes (33). Thanks are due to David Pinniger of the Pest Infestation Laboratory, Slough who kindly read and commented on the chapter dealing with pests and treatments (4). We must also thank Barbara Parris (Auckland, New Zealand) and Father K.M. Matthew (Tiruchirapali, India), for providing additional references.

Diane Bridson and Leonard Forman

CONTENTS

ADDITIONAL TECHNIQUES

COLLECTING

THE HERBARIUM IN A WIDER CONTEXT

INTRODUCTION

Until the present century, the science of botany in Europe has been dominated by plant taxonomic studies. This was to a great extent the result of the colonial expansion by European nations in the past five hundred years which had the effect of bringing to the attention of European naturalists an ever-wider, more complex and largely unknown range of plants and animals, mainly from the tropics.

In the twentieth century, botanical science has developed rather differently, with a great expansion of molecular, cell and physiological studies, which have displaced taxonomy from its former primary position. This has led to a view among many scientists that taxonomy is an out-dated, minor discipline of little importance in the modern world. In our own time, this opinion has been shown to be a serious error of judgement. The drastic degradation of the environment which is apparent throughout the world today has demonstrated how little we know about the plant diversity which is so fundamental to successful human existence. To a very large extent, plants *are* our environment. Regions where plants are abundant tend to be places where man can succeed, and where they are absent man has little hope of survival. Knowledge of plant diversity is essential in the present struggle to reclaim deserts and degraded landscapes, breed pest resistance into our crops, and find new sources of energy, food, medicines and useful materials. Many scientific disciplines are involved in this work, including ecology, conservation, physiology, plant breeding, pharmacology, biochemistry, ethnobotany, materials technology, agronomy, plant pathology and many others. Constantly, and with increasing urgency, questions such as the following, which only the taxonomist can answer are being asked about plants:

- How can they be recognised? (IDENTIFICATION)
- What should they be called in order that information about them can be freely exchanged without ambiguity? (NOMENCLATURE)
- What are their closest relatives? Are there any other plants likely to have similar properties or compatibile genetic systems? (CLASSIFICATION)
- Where do they grow? (DISTRIBUTION)
- In what kind of habitat do they grow? (ECOLOGY)
- Have they any useful properties? (USES)

The role of taxonomy today is to be able to give quick and accurate answers to these questions. It is a vital role, though not perhaps a spectacular one. Taxonomists must carefully amass their data, sift out the bad, refine the good, and provide an efficient service to science as a whole. In this task the herbarium is the most essential working tool for the taxonomist.

Our capacity to answer questions from the wider world of science and technology depends to a very large extent on the quality and completeness of representation of our herbaria. It therefore follows that all aspects of herbarium work must be considered fundamental to the good health of our profession. In short, without herbaria there can be no taxonomy.

1. WHAT IS TAXONOMY?

The introductory statement of Lawrence (1951) in his 'Taxonomy of Vascular Plants' will serve us well as a definition of the term taxonomy: 'It is a science that includes identification, nomenclature and classification of objects, and is usually restricted to objects of biological origin; when limited to plants, it is often referred to as systematic botany.' One can be more restrictive and limit the term taxonomy to the study of the principles underlying a system of classification, but in the present context the Lawrence definition can be used.

The three elements of this definition can be further explained as follows (adapted from Lawrence):

1. Identification is the determination of a plant or taxon (such as a species or subspecies, see **11. Plant names**) as being identical with or similar to another and already known element. In some instances, the plant may be found to be new to science.

2. Nomenclature is concerned with the determination of the correct scientific name of a known plant according to a nomenclatural system; that is, it may have a handle by which it can be referred. This naming is regulated by internationally accepted rules laid down in the International Code of Botanical Nomenclature.

3. Classification is the placing of a plant (or group of plants) in groups or taxa, which are referred to various categories according to a particular plan or sequence; e.g. every species is classified as a member of a particular genus, every genus belongs to a particular family etc.

With a full understanding of the above definition we can appreciate the role and value of taxonomy.

We take for granted that we refer to one another by name, e.g. 'John Smith', and therefore we have a reference point for relating information to a source. Our names are given to us in a structured way. For example, in many countries our surname is the same as our father's surname and our first name is chosen for us by our parents. Similarly, many everyday objects have compound names which immediately enable us to distinguish between different kinds of similar objects, e.g. when referring to spoons we speak of an egg-spoon, tea-spoon, dessert-spoon and table-spoon. This two-part name or binomial is very similar in structure to that given to a plant by the botanist, who provides a generic name and a specific epithet, which together form a scientific name for the plant being identified.

Many plants have local names and these are often useful within a local community but outside the region the name becomes meaningless and therefore the related data are no longer accessible. A classic example in the UK is the local name 'Bluebell' which in Scotland is a species of *Campanula* but in England is a member of the Hyacinthaceae. Hence the universal Latin name

Hyacinthoides nonscripta for the English Bluebell, prepared in conformity with the International Code of Botanical Nomenclature, provides the unambiguous nomenclature which is required.

By arranging the entities which the scientific names represent in various systematic classifications, we have the other side of the botanist's work – the creation of a taxonomy or a botanical system. The advantage of arranging the species in genera and the genera in families etc., is that it allows us to deduce expected characteristics of a given taxon (such as a species) from a knowledge of its close relatives. Taxonomy therefore has a predictive value. The taxonomic structure thus allows us not only to pick out plants with a similar or related floral or vegetative structure, but also to associate data on chemical, anatomical, cytological features, uses and habitat tolerance; in fact the whole of the characteristics associated with any one species. These data are meaningful only if they can be associated with a name to which other workers in different disciplines can relate.

Therefore, one may see taxonomy as the frame or base, by means of which all other plant data can be related.

REFERENCES

Jeffrey (1982)
Lawrence (1951)

2. THE DEVELOPMENT, PURPOSE AND TYPES OF HERBARIA

THE DEVELOPMENT OF HERBARIA

The word Herbarium in its original sense referred to a book about medicinal plants. Tournefort (c. 1700) used the term for a collection of dried plants; this usage was taken up by Linnaeus under whose influence it superseded earlier terms such as 'hortus siccus' (Stearn 1957).

Luca Ghini (1490?–1556), a Professor of Botany at the University of Bologna, Italy, is thought to have been the first person to dry plants under pressure and mount them on paper to serve as a lasting record (Arber 1938). This practice spread throughout Europe and by the time of Linnaeus (1707–1778) the herbarium technique was well known. Many early herbaria were bound into book-like volumes rather than being kept as separate sheets. At first, herbaria were mostly privately owned, but the practice of depositing specimens in established collections and exchanging specimens or selling collections was a common practice by Linnaeus's time. See DeWolf 1968: 70–71; Radford et al. 1974: 751–752.

THE PURPOSE OF A HERBARIUM

The main questions which taxonomy sets out to answer have been outlined in the **Introduction**. The resolution of these questions requires examination of the plants themselves, but it is impractical to obtain more than a few of the answers from plants in a living condition. Taxonomic research therefore relies upon a collection of preserved plants built up over a long period of time – *the herbarium*.

Botanists are fortunate in that the standard method of preparation – mounting on sheets – enables their specimens to be treated like cards in a filing system. Only a limited amount of awkward material needs to be treated differently in special ancillary collections with their associated problems of cross-referencing. The herbarium is thus a simple idea, though capable of considerable elaboration as the purposes for which it is to be used expand in scope.

These purposes are set out below, and they make increasing demands on the botanical knowledge and organizational skill of the curator.

- A store of reference material. This requires adequate arrangements for the preservation of specimens, and a simple form of indexing (such as alphabetical) that enables them to be retrieved quickly.

- A means of identification, by matching unnamed plants with named specimens in the collections. To do this effectively the specimens must be arranged in a way that bears some relation to their overall similarities. Alphabetical arrangements, for example, must therefore be replaced by taxonomic arrangements which reflect relationships.

- An arbiter of correct names. Printed Floras soon become out of date, and it is up to the herbarium to maintain nomenclatural standards. This entails keeping the names in line with current revisionary work, maintaining a type collection, and organizing exchanges of specimens with other institutions.

- A comprehensive data-bank. Ideally the collections should fully represent the diversity and distribution of the region's vegetation, and the head of department should organize collecting expeditions to fill gaps in the coverage. Many herbaria will require a geographical arrangement superimposed on the systematic one.

DIFFERENT TYPES OF HERBARIA AND THEIR FUNCTION

The size of the herbarium and the type of work it undertakes will, to a large extent, dictate the methods of arrangement and curation needed to run it as an efficient 'machine'. Turrill (1964) suggested considering two main types of herbaria, the general and the special (further subdivided). For the present purposes four main types will be considered (general, national, local and special), but these grade into one another.

General (or International) Herbaria – these are very large herbaria often with 4 million or more specimens and a global representation of as near a comprehensive range of taxa as possible. Most general herbaria were founded early in the history of formal taxonomy and have grown to their present size over the years. Because they are rich in types and other historical specimens they attract many visitors and loan requests. The functions of general herbaria include:

- Broadscale studies of families and above.
- Production of generic monographs (with special attention to generic limits); major floras (covering several countries); national and local floras; check-lists.
- Services include: loans; providing facilities for visiting botanists; identifying specimens (especially noting new taxa) and dispatching determination lists; distributing duplicates.

National or (Regional) Herbaria – geographically these cover the country concerned and neighbouring or phytogeographically similar areas. As far as possible all taxa relevant to the area should be represented. National herbaria may be relatively old or modern (depending on the history of the country). Type material is often well represented especially among the more recently described taxa. The functions of National herbaria include:

- Contributions to major floras (covering several countries).
- Production of national and local floras; check-lists.
- Services include: loans; providing facilities for visiting botanists; identifying specimens relevant to the country and dispatching determination lists;

collecting material from the field and distributing duplicates; providing material for ancillary disciplines, (e.g. anatomy, cytology, chemistry), especially material freshly collected for the purpose.

Local Herbaria – these deal with a region within a country such as State, County or District or even a much smaller area such as game park or nature reserve. Local herbaria usually have a relatively short history and contain few (if any) type specimens. All taxa relevant to the area should be represented, but it is not necessary to keep large numbers of each taxon. The functions of a local herbarium include:

- Contributions to national floras.
- Production of local floras and check-lists.
- Services include: identifying specimens relevant to the area and compiling determination lists; collecting material from the field and distributing duplicates; collecting material from the field for ancillary disciplines.

Special Herbaria – these are often but not always small and have limited scope or a specific purpose; there are several types of special herbaria depending on function:

- *Historical herbaria*: these may be kept as separate herbaria within a general herbarium (e.g. Wallich Herbarium at Kew or De Candolle Herbarium at Geneva) or belong to a separate institution (e.g. Linnean Society or universities, museums or monasteries). They are usually arranged in their original sequence (not updated to current taxonomic opinion) and have restrictions governing consultation and loans. Because they are rich in type specimens the more important historical herbaria have been put on microfiche to help overcome the problem of accessibility.

- *Herbaria of limited scope*: these may be either limited taxonomically (e.g. cryptogamic herbaria) or ecologically (e.g. forest herbaria). Many herbaria in this category are of considerable size and importance and should be considered with national herbaria. They are often separately housed within general herbaria or other institutions such as universities or museums.

- *Teaching herbaria*: these are housed in universities, with more modest herbaria in colleges and schools (some university herbaria rank as major and should be considered with national or local herbaria). Teaching herbaria should contain specimens to illustrate morphological structures, the types of plants representative of communities encountered in field studies, examples of both economic species and species of locally grown crops, and a series to illustrate families and genera for taxonomic instruction.

- *Job-related herbaria*: these could include collections of weed species for agriculturists or honey-bee plants for bee keepers, for example. They should contain good examples of all relevant taxa but multiple collections of the same taxon are not desirable. Collections of cultivated plants can be useful adjuncts to botanical gardens, arboreta, nurseries or agricultural stations, especially if exotic species are grown. Differences in cultivated forms (cultivars) of ornamental species can be subtle and lost in the dried specimens; these should be supplemented with records of the exact flower colours and photographs to show habit and details of form.

— *Herbaria for special research programmes*: voucher specimens for the following: anatomy, cytology, chemical studies, ecological surveys, host plants of insect or fungal pests, animal food-plants. Voucher specimens are often housed in general or national herbaria but they are frequently unwelcome as they can lack quality, are often of common taxa and take up valuable space. If housed in the university department sponsoring the research, the researcher has the convenience of direct access to the vouchers. However, a decision should be taken as to whether the collection is to be permanently or temporarily retained.

REFERENCES

Arber (1938)
Beaman (1965)
Brenan (1968)
Cronquist (1968)
DeWolf: 70–71 (1968)
Jain & Rao: 8–14 (1977)
Radford et al.: 751–756 (1974)
Shetler (1969)
Stearn (1957)
Subramanyam & Sreemadharan (1970)
Turrill (1964)

3. THE HERBARIUM BUILDING AND SPECIMEN STORAGE

A herbarium building is basically a storehouse of scientific botanical specimens. The quality of the collection and in turn the standard of work that can be undertaken in using it must depend, in part at least, on the type of storage and the environment in which the specimens are kept.

Often, by historical accident or financial constraint, storage facilities are restricted to what is available, and thus may not be the best. It may, however be helpful to consider the main requirements.

Location

If there is freedom to choose:

- Avoid areas liable to flooding or adjacent to a flammable building or vegetation.
- Avoid soft ground areas liable to subsidence and hillside slippage.
- Construct the building in or near a living collection, such as a botanic garden, to facilitate study in association with live plants.
- Aim for easy access to accommodation both for visitors and staff.
- If possible avoid placing it under the flight path near an airport where there is an increased danger of plane crashes and constant noise (e.g. Kew!).

Construction

- Preferably, a herbarium building should be purpose-built but this may not be possible. Most importantly it must enable the collections to be housed in dry, safe storage and facilitate a rational arrangement of the specimens.
- It should be fire-proof, waterproof and, where appropriate, earthquake-proof. In the tropics timber construction should be avoided as it can be subject to attack by termites. Modern cellular walls should be avoided as they can harbour pests such as cockroaches.
- If a new herbarium is planned, allow for generous future expansion. It is always cheaper to provide more space in the building initially than to create it later. Alternatively, the design should facilitate future extension. If the addition of more floors is anticipated, consider the strength of the foundations in the initial plan. The ceilings should be relatively low so that cabinets and shelving can be kept to a comfortable height.
- If re-developing an older building to house a herbarium it may be necessary to construct intermediate floors (or mezzanines) to ensure that cabinets or box stacks are never too high.

– Certain functions should be carried out away from the main herbarium. If possible separate buildings should be provided or self-contained areas isolated within the main building, see below, *Facilities for specimen reception and decontamination, Poisoning room* and *Drying facilities.*

Essential facilities

Apart from the herbarium the main building should house the following facilities:

– The appropriate facilities for staff. It is particularly important to provide some rest room facility, because much of the work in the herbarium involves standing or walking about. Separate rooms in which food may be consumed are also needed.

– Ground floor reception area for visitors near and in sight of the main entrance.

– Office space: general offices for common service functions (e.g. secretarial and centralised records) and private offices for the Director and other appropriate staff members.

– Work stations for staff and visitors: these may either be within the main herbarium or in separate rooms.

– A mounting room.

– A separate room for spirit (wet) collections.

– A library, or provision for the essential literature if the main library is housed elsewhere (e.g. in a University).

The following facilities should by preference be housed separately from the main herbarium:

– *Facilities for specimen reception and decontamination.* Preferably the reception area for specimens should be in a separate building, or at least in a room totally isolated from the main herbarium. There should be adequate room for the specimens to be inspected for pests and treated without risk to the main herbarium. Allow for plenty of reception shelving; the environmental conditions should resemble those of the main herbarium as far as possible.

Accommodation must also be provided for the equipment needed for decontamination. Special ventilation is necessary if chemical methods are used. (See below, *Internal design* and **4. Pests and treatments**).

– *Poisoning room.* Apart from decontamination there may be a need to poison specimens to prevent future attack (see. **4. Pests and treatments**). A separate room must be provided for this process with its own ventilation direct to the outside air, and not via air-conditioning if this is installed. A fume-cupboard should be installed, as well as a wash basin.

– *Drying facilities.* A facility for pressing and drying plant specimens is usually required. Preferably a separate room should be provided to house the drying unit, together with bench space on which the specimens can be prepared and the presses changed, and racks to hold extra drying paper.

Drying units range from specially designed electric ovens (or drying cabinets) to simple box-like structures over a safe heat source (e.g. light bulbs), similar in principle to field driers. (See also p. 213). Electrical units should be installed by a qualified electrician (or engineer familiar with the wiring of the building). The oven must have thermostatic controls and ideally be maintained at 50°C. It is important to install a circulating and extracting fan and a flue (or chimney) leading the moist air to the exterior of the building. The shelves should be variable in position and made of wire mesh. See McClean & Storey 1930; Schnell 1960: 39–41; Smith 1946; Gates 1950.

Internal design

Temperature and humidity

The ideal conditions for paper conservation are 20–23°C at ± 55% humidity, while dried plants are better preserved in somewhat drier conditions. Low humidity reduces the risk of fungal infection and the risk of breeding colonies of insect pests becoming established.

If possible, the temperature should be maintained at about 20–23°C, and the humidity at 40–60%; these conditions are also ideal for staff efficiency. Depending on the local climate, it may be necessary to install central heating and/or air-conditioning to control the temperature throughout the year, and possibly de-humidifiers to reduce humidity. In tropical regions, if de-humidifiers cannot be obtained, locating the herbarium on the upper floors may be of help in reducing the humidity. Bear in mind that in tropical regions air-conditioning should not be set at too low a temperature (i.e. not below 20–21°C) because the sharp contrast with the outside temperature is unpleasant for people when leaving the building. See CCI 1990: TB1; DePew 1991: 45–59; Michalski 1992.

Ventilation

Adequate ventilation is essential and it is important to aim for the renewal of air, while avoiding excessive loss of pest-preventative substances or the entry of humid air or dust. Dampness is less likely to occur if ventilation is adequate.

Good ventilation contributes to staff comfort, while lack of ventilation can be a health hazard. In conditions where the atmosphere within the herbarium rooms has a high concentration of insecticidal, fungicidal or insect-repellant vapour (e.g. naphthalene), staff should work in separate rooms. A fume-cupboard should be installed if the staff are to handle any volatile substances, e.g. in the spirit collection (see **13. Ancillary collections**, *Spirit collection*).

To prevent the entry of insect pests, windows and external doors should be fitted with draught excluders and, in the tropics, with insect-proof screens.

Lighting

It is important to ensure adequate lighting. Natural lighting is the best, but in addition try to install fluorescent-tube lighting, positioned carefully to minimise shadow areas. It will also be necessary to make provision for desk lamps, as many people find it uncomfortable to read and write by fluorescent light (unless of the very high frequency type).

Power points

Now that computers occupy a central role in herbaria it is important to ensure that sufficient power points are installed to accommodate them as well as other electrical equipment (e.g. microscope lights, boiling rings etc.). If a mainframe computer is used, sufficient terminals will be needed. It will also be necessary to install telephone points.

Fittings and fixtures

Floor surfaces should be easily cleaned and maintained but not slippery. Loose-fitting floor coverings such as carpets and linoleum should be avoided as pests can shelter beneath them. Fitted tiles or wood blocks are better.

Fitted furnishings must be installed so that 'dead spaces' in which insect pests could survive are avoided. Sufficient space must be left beneath, above and in the corners to allow easy access for cleaning and regular inspection for evidence of pests. Otherwise the fittings must be tightly abutted at floors, ceilings and corners to exclude insects.

Work surfaces should be reasonably close to the herbarium storage units, and should be large enough to spread out specimens easily when sorting, and be of a comfortable height when standing. They can usefully be designed to incorporate cabinet space beneath them. It is also necessary to provide work space at sitting height.

Storage units

It is important to consider whether storage in boxes is preferable to cabinet storage. If the design of the building prevents storage adjacent to the working surface, or a part of the collections has to be kept separately, the specimens may be some distance from the working area. In these circumstances they are best transported in boxes.

- *Boxes*. The best boxes are made from cloth-covered hardboard and are specially constructed to have a drop-flap front with a label-holding slot as well as a lid-top (full or partial). However, in some tropical herbaria it may be better to house the specimens in metal boxes which are more insect-proof. It may be possible to obtain suitable ready-made boxes, e.g. for office use, but it is important to avoid any that do not have a drop-flap front, as the specimens can be damaged if removed from above. (See also **17. Curation of special groups**, *Palm-boxes)*.

- *Cabinets*. The material from which herbarium cabinets are made should be considered. It is worth noting that although wood is flammable and metal is not, in the event of a fire specimens within a metal cabinet will be quickly charred, while those in a wooden cabinet may be insulated from the heat for some considerable time. Metal cabinets are, however, more insect-proof and this may be a deciding point in tropical herbaria. Cabinets made of some tropical woods (e.g. *Cedrela odorata)* may stain the specimens. The use of foil-backed shelf lining paper will overcome this problem.

  Cabinets should preferably have shelves which are at least 2 cm. deeper and wider than the herbarium sheets and approximately 15 cm. apart. The doors of the cabinets should be close-fitting with a dust-proofing strip of, for example, rubber or felt, and the catches should fasten tightly. These points are important for the exclusion of insect pests. A programme of routine maintenance may be necessary to ensure a continuing good fit of cabinet doors. Some designs of herbarium cabinets incorporate a slot on the inside of the door to contain insect repellent or insecticidal substances. See Jain & Rao 1977; Mori et al. 1989.

- *Mobile storage units (Compactors)*. Where space is limited, the cabinets or shelves which carry boxes can be installed as mobile storage units. These are blocks of cabinets or shelves mounted on runners, enabling nearly the whole floorspace to be filled. The units can be closed together or separated where required, so leaving in each block at least one open gap for access to the specimens. The units may be hand-operated or motorized, but the latter can break down or the electrical supply can fail. A disadvantage of this system in a busy herbarium is that in each block of units usually only a single open gap is available, thus allowing only one person to work at a time. See Davidse 1975; Jain & Rao 1977; Touw & Kores 1984; Cholewa & Brown 1985.

Fire precautions

A number of important herbaria have been destroyed by fire with the loss of valuable scientific material, the product of many years of work. Strict fire precautions must be enforced, including a complete ban on smoking and the use of naked flames. All rooms, corridors and stairways must be separated by adequate fire-doors, and regularly maintained fire-extinguishers must be placed in all areas of the building. Advice must be sought, probably from the local fire-brigade, concerning the installation of fire-hydrants and emergency fire-hoses. However, the possible damage to the collections by water, in the event of an emergency should be considered. A fire-alarm system must be installed and a plan made for the procedures to be followed if a fire does occur. All staff must be familiar with the fire-drill and occasional fire-practices should be held.

REFERENCES

CCI, TB1 (1990)
Cholewa & Brown (1985)
Davidse (1975)
DePew (1991)
Fosberg & Sachet: 59–64 (1977)
Gates (1950)
Jain & Rao: 67–68 (1977)
McClean & Storey (1930)
Michalski (1992)
Mori et al. (1989)
Schnell (1960)
Smith (1946)
Touw & Kores (1984)
Womersley: 75–76 (1981)

4. PESTS AND TREATMENTS

Long-term preservation of herbarium specimens involves constant vigilance to protect them from damage by pests. With few exceptions, the most serious threat to herbarium material is from insects, and this threat is highest in tropical and sub-tropical herbaria. In tropical regions the high temperature and humidity levels allow pests to grow and multiply rapidly, and the numerous indigenous pests easily gain access to the herbarium. In temperate regions most pests do not live outside heated buildings and so reinfestation is a less frequent occurrence. However, some pests typical of temperate conditions (e.g. silverfish) have become established in modern tropical herbaria where air-conditioning has created a suitable climate.

A recent survey entitled 'Pest Control in Herbaria' is given by Hall 1988 and there is a guide to 'Insect Pests in Museums' by Pinniger 1990.

TYPES OF PEST

Insect pests

It is important to identify accurately any insects suspected to be herbarium pests. If possible, take specimens at all developmental stages to an entomologist; otherwise, try a manual, e.g. Zycherman & Schrock 1988. Learn as much as you can about the pest's life-cycle and requirements: does it fly, what does it eat, where does it live and under what conditions of temperature and humidity does it thrive and breed? Usually it is the larval stage that causes most damage.

Insects can be divided into three groups depending on the kind of damage they cause.

The first group is by far the most destructive as the pests *feed directly on the preserved plant* material as well as on paper and glue, thus severely damaging the specimens. The two causing the greatest problems are:

- Drug-store, biscuit beetle or 'Herbarium' beetle – *Stegobium paniceum*. (See fig. 1, A), the commonest pest in temperate and highland tropical herbaria. It can survive and breed on drugs and spices that would be toxic to other insects.

- Cigarette or tobacco beetle – *Lasioderma serricorne*. (See fig. 1, B), the commonest pest in tropical herbaria.

Both the above beetles fly actively in the adult stage and can therefore enter herbarium buildings through unprotected open windows.

The second group are *scavengers* and will feed on mould or detritus. If the populations are low they will constitute a nuisance rather than a hazard, but if their numbers build up, serious damage can result. Pests falling into this category include:

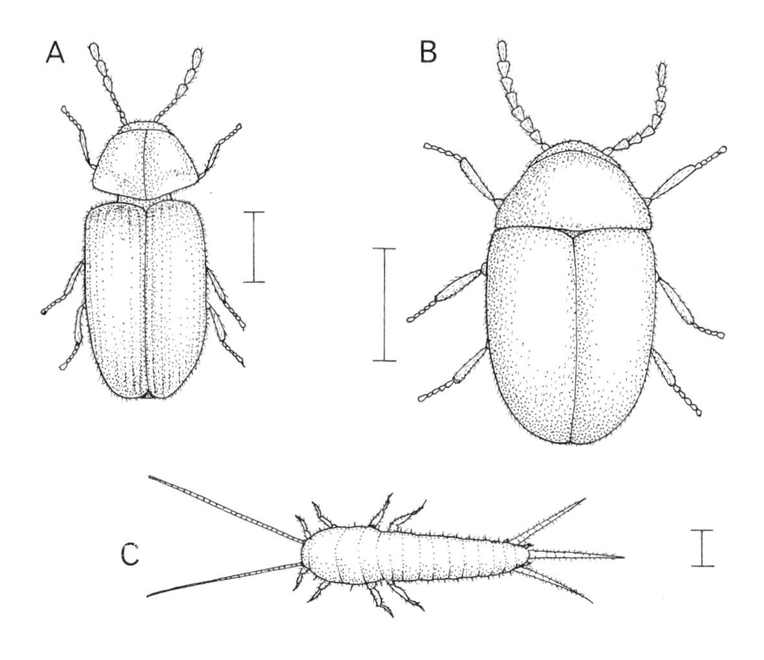

FIG. 1 — **A** *Stegobium paniceum*; **B** *Lasioderma serricorne*; **C** *Lepisma saccharina*. All scale lines = 1 mm. N.B. in life the head and legs of A and B are usually tucked under the body.

- Spider beetles, particularly the Australian spider beetle – *Ptinus tectus,* small spider-like beetles, common in birds' nests and general debris. Can complete development below 10°C and eggs are said to survive 0°C. Will make holes in paper and specimens.

- *Cartodere filum* (a small 1.2–1.6 mm. long, brown slender flat beetle) – frequently occurs in herbaria, in both tropical and temperate regions, especially on mouldy specimens. Usually it will only eat the mould rather than the specimens, but in very large numbers it has been known to cause damage, particularly to mouldy fruit in carpological boxes. This pest can be a more serious problem in collections of fungi.

- Silverfish or fishmoths – *Lepisma saccharina* (See figs. 1, C & 2) – feed on paper and glue. The surface of the paper together with the writing can be grazed off. Silverfish are wingless, active at night and capable of moving from room to room. They are associated with damp conditions and require local humidity above 75–80% to breed.

- Booklice (Psocids) – *Liposcelis* (several species) – are 1–2 mm. long, all female, wingless and relatively fast moving. Populations can increase very rapidly at temperatures of 25°C or over and at relative humidities as low as 60% but thrive at higher humidities. Booklice feed mainly on surface moulds but they also eat paper and glues and have been reported to damage herbarium specimens of Compositae (Asteraceae), Cruciferae (Brassicaceae) and petaloid Monocotyledons, and to eat pollen.

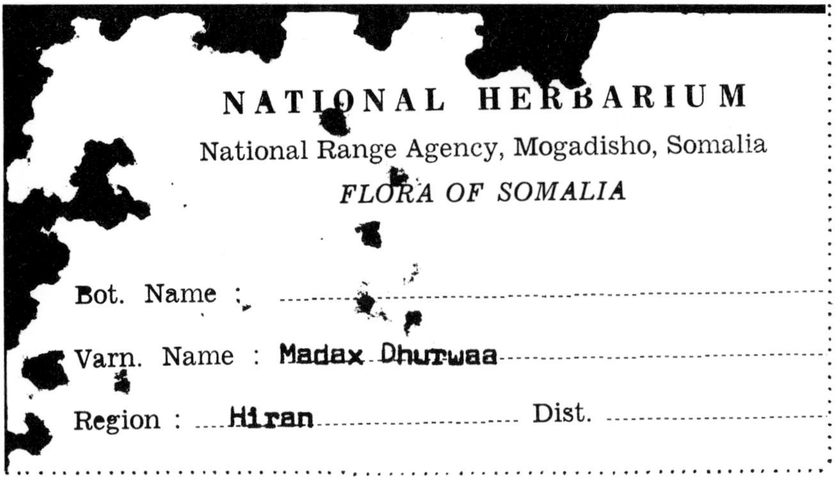

NATIONAL HERBARIUM

National Range Agency, Mogadisho, Somalia

FLORA OF SOMALIA

Bot. Name : ....

Varn. Name : Madax Dhurwaa..

Region : ....Hiran.......................... Dist. ....

FIG. 2 — Herbarium label damaged by silverfish.

- Cockroaches – especially the German cockroach – *Blattella germanica*, 12–15 mm. long and the American cockroach *Periplaneta americana*, 30–40 mm. long. Cockroaches do little direct damage to herbarium specimens but they may eat glue and damage leather and cloth bindings of books etc. Faecal material may make specimens unsightly and stain the paper. They are often associated with catering areas and inhabit wall-cavities, heating ducts etc. They are nocturnal and prefer moist warm conditions and are known to carry a range of diseases which could present a health hazard to staff, especially if food is consumed on the premises. The help of a pest control specialist will be needed for their eradication.

- Ants including Pharaoh's ant – *Monomorium pharaonis* cause considerable nuisance if they invade herbaria in numbers but they seldom damage specimens. They may become established in heating systems in temperate herbaria.

- Carpet beetles and moths usually feed on animal products such as fur, feathers, wool, silk and leather. They are not usually a problem in the herbarium but can damage library material and ruin leather bookbindings. They can also make holes in paper. One exception is the Varied carpet beetle – *Anthrenus verbasci* which has been reported to feed on seeds and pollen, especially in members of the Compositae (Asteraceae).

The third group are the *woodborers*, especially those that live on seasoned timber; they can cause damage to wood collections and to the fabric of the building. Some of the most destructive are:

- Drywood termites – *Cryptotermes spp.* and *Kalotermes spp.* A tropical group that can cause extensive damage before detection as their nests are constructed entirely within timber and they do not forage widely. They will also attack and destroy other cellulosic material such as herbarium sheets and books.

- Furniture beetle or woodworm – *Anobium punctatum*. Widespread in temperate countries.

- Death-watch beetle – *Xestobium rufovillosum*. Confined to old hardwood timbers in historical buildings.

- House longhorn beetle – *Hylotrupes bajulus*. Attacks structural timbers (mostly softwoods) in many countries.

- Powder-post beetles – *Lyctus spp*. Will attack temperate deciduous hardwoods, tropical hardwoods and bamboo.

- Bostrychid beetles – *Dinoderus spp*. etc. A large tropical group which will attack hardwoods and bamboo.

Non-insect pests

The only other significant pests are:

- Mites, which do not generally harm the collections severely but may attack fungal herbarium material. Their presence often gives an indication of high moisture levels and poor storage conditions. Some mites can cause allergic reactions in staff.

- Moulds and mildew, a constant threat to material stored in damp conditions or in areas of high humidity.

- Mice and rats using herbarium specimens for nest material or (occasionally) chewing into cabinets, e.g. containing specimens of cereals such as *Sorghum*.

DETECTION

In a working herbarium a constant look-out should be kept for insect and any other kinds of damage during routine activity. However, it is vital that this is supplemented by *regular* checks for infestation. A detailed plan of the building should be consulted and a note made of any separate or ancillary collections, especially if infrequently consulted – e.g. historical, carpological and wood collections. It is also important to remember to search any specimens housed in store rooms, e.g. duplicates awaiting dispatch or specimens held in botanists' individual offices. Some pests may be seasonal and it is important that searches are timed to coincide with their optimum activity. Ideally the date and details of the search should be recorded in a *log book*. The most useful ways of checking for pest infestation are:

Visual inspection

Most insect damage is caused by the larval stage; evidence of the presence of larvae is given by deposits of fine granular droppings (frass), missing parts of plants, holes in leaves, and crawling or dead adults; or in wood a fine powder-like dust and exit holes. It is important that old debris and frass are removed so that any new deposits are immediately apparent.

Check for evidence in certain plant families to which insect pests are known to be naturally attracted – Compositae (Asteraceae), Cruciferae (Brassicaceae) and Capparaceae, and petaloid monocotyledons are the most frequently attacked by several common pests; if *Stegobium paniceum* is present, check Compositae, Umbelliferae, Ericaceae, and plants that contain latex, e.g. Apocynaceae and Asclepiadaceae. Flowers and young stems are particularly vulnerable.

The use of traps

Strategically placed traps can be used to monitor the presence of adult insects. This does not dispense with visual inspection but supplements it. Traps designed for both crawling and flying insects are available (sticky traps and funnel traps). One designed especially for cockroaches also traps other pests including booklice and furniture beetles (Pinniger 1991). A *Lasioderma* trap has been produced for the tobacco industry which lures male insects to the sticky trap by a pheromone (the attractive scent of the female). A similar trap has also been developed for *Stegobium*. All traps are non-toxic and can give an earlier warning of infestation than visual inspection alone.

Some tropical herbaria use lamps with a sticky surround to attract and trap night-flying insects. It is important that such lamps are regularly checked for safety and they should on no account be used if there is the slightest danger of fire risk. It must be remembered that crawling insects may not be caught.

PREVENTION OF INFESTATION

The methods of pest control can be divided into chemical and physical. Chemical methods must be used with *great caution* because most insecticides can:

- directly harm the health of the staff
- build up to harmful levels in the environment
- fail to kill pests, as many insects can develop tolerance to specific toxic chemicals
- be very costly

Physical methods, although safe to people, non-polluting and less expensive, can in some circumstances damage the specimens. Since there is no single ideal method of pest control, a combination of approaches, or an **integrated pest control** method, should be used to give the best possible protection. One staff member should be in charge of the co-ordination of the pest control strategy but it is important that *all* staff (including cleaners) are fully informed of its aims and methods and most importantly of any hazards. A written account, in the form of a log or diary, should be kept of all control procedures in case the officer concerned leaves the staff.

Exclusion of pests. Minimize insect access to the building and collections by:

– fitting fine screens and well-fitting draught excluders to opening windows and external doors, ceiling ductwork and wiring gaps.

– making sure all herbarium cabinet doors are kept tightly shut and fitting them with draught-proof seals if possible.

– **NOT** bringing specimens into the herbarium *until* they have undergone decontamination (see below, *Decontamination methods*).

– It is advisable to subject new stocks of corrugated cardboard to a decontamination process before they enter the herbarium building. They could harbour pests such as booklice and spider beetles if previous storage has not been in ideal conditions.

– In tropical and subtropical herbaria it is advisable not to bring pot plants or cut flowers into the herbarium and to remove vegetation which could harbour pests growing close to the building.

Environmental control (see also **3. The herbarium building and specimen storage**). The following are recommended:

– The temperature should be kept at 20–23°C and the humidity at 40–60%.

– The furniture should be fitted so as to avoid 'dead spaces' (i.e. inaccessible gaps below cabinets etc.) that cannot be easily cleaned.

– Birds' nests and debris must be removed from roofs and lofts.

– The interior should be kept clean; dust should never be allowed to accumulate and food should not be consumed within the herbarium working area.

– Heating and air-conditioning ducts should be kept clean.

Regular fumigation of herbarium rooms or cabinets. Several gases such as methyl bromide, phosphine, ethylene oxide and less commonly carbon disulphide, carbon tetrachloride, hydrogen cyanide, ethylene dichloride and sulfuryl fluoride (Vikane) have been used. This method is still used in some areas with very acute pest problems (see Hall, 1988); it is, however, undesirable for the following reasons:

- Effectiveness depends on regular repeat treatment, e.g. annually in tropical regions.
- It does not necessarily kill the eggs or pupae of the insects (except in a vacuum chamber).
- The gases are very poisonous and fumigation should *only* be carried out by people who have been trained in fumigation techniques.
- The gases may have harmful environmental effects.

Insect-repellants, such as naphthalene or paradichlorobenzene (PDB), are sometimes placed in small quantities in herbarium cabinets but are *not* recommended for health reasons. Furthermore, to be effective for beetle pests the concentration in the cupboards needs to be higher than for moths; this in turn increases the health risk. Do not use both naphthalene and PDP together in warm climates because an oily mess can result (Womersley 1981: 73). PDB can soften some plastic mountants (Croat 1978: 206; Womersley 1981: 73). It is possible that essential oils, such as lavender, applied on wads of cotton wool, could be used as insect-repellants. There are however, no experimental data available.

Poisoning to make specimens unpalatable to insects and mites:

Poisonous substances applied in solution may be used to act as a semipermanent deterrent to insect infestation. The solution may be applied by immersion (unmounted specimens) or painted or sprayed on (mounted specimens). It is advisable to label or to stamp the specimens to indicate they have been treated in this way.

- *Mercuric chloride** (*corrosive sublimate*) has been used to protect specimens from insect attack, but this is a *very dangerous substance to handle* and in some countries legislation may prevent or control its use. It may damage and blacken certain types of paper and can leave fine crystalline deposits resembling hairs on the specimens (Clark 1988: 6; Fosberg & Sachet 1965: 66; Womersley 1981: 73). It is **not** recommended.
- *Lauryl pentachlorophenate* (LPCP) (also known as dodecanoic acid, *Mystox* or phenyl pentachlorophenate (or pentachlorophenyl ester) = PPCL). This is used in solution: 3.75% in white spirit, in 5% 'Varsol' or in 'Shell Odourless Carrier'. It has been used in both temperate and tropical herbaria. It was

* Some authors, e.g. Womersley (1981), call a solution containing mercuric chloride 'Kew Mixture'; it should **not** be confused with Kew Mixture as used for a liquid preservative.

used first at the Natural History Museum, London in 1962. When used correctly LPCP is generally said to be safe. It does not damage clothing or discolour paper, nor affect printing, permanent ink or pencil. It apparently leaves plant surface features intact (fide Fosberg & Sachet 1965: 67) although there seem to be no recently published data. It does, however, normally require bundles of unmounted specimens to be totally immersed in the solution, and this causes blurring of typescript, carbon copies and ball-point and washable inks. Absorbent heavy-duty wrappings and glossy printed paper which will leave ink marking on anything in contact with it should be removed before the bundle is dipped, as should also grasses and ferns, which do not need treatment (see Fosberg & Sachet 1965: 66; Womersley 1981: 79; Mori et al. 1989: 41–43). LPCP does not confer as long-lasting an immunity as mercuric chloride and there is some evidence that booklice (*Psocids)* and tobacco beetles can live on specimens poisoned with it (see Guillarmod 1976; Womersley 1981: 71). Reapplication may be necessary; LPCP can be applied to mounted specimens by spraying or with a small brush (see Whitmore & Fosberg 1965; Lundell & Kirkham 1966).

– *Sodium pentachlorophenate* has been used in the Philippines (see Eusebio & Stern 1964).

Desiccant dusts made from fine particles of diatomaceous earth or silica act by damaging the surface layers of wax and cuticle of the insects so that they lose water and die. Although not suitable for direct use on the specimens, the dusts can be placed beneath cabinets or in other dead spaces. Some brands are mixed with natural pyrethrins (see below) for rapid killing.

Insect traps (see above), can help to control small infestations by reducing the population level or by interrupting the breeding cycle.

Insecticides can be used both to treat small local infestations and to prevent build up of populations. There are many different chemical types of insecticides. Most are harmful to people, and others are persistent and cause environmental pollution. Because of this many are strictly controlled or prohibited in certain countries. It is important that the manufacturer's instructions are precisely followed; it is now an offence under the UK Pesticide Regulations not to do so. It is important to remember that insecticides *must* be used in sufficient concentration to kill the pests. If less is used as an economy, the insects may develop resistance and the problem then escalates. For further information on the properties of insecticides see Hall 1988, Pinniger 1990; Zycherman & Schrock 1988. The most frequently encountered insecticides are:

– *Pyrethrins* (manufactured from the flowers of *Chrysanthemum cinerariifolium)* are the safest insecticides. They are not very toxic to people and are rapidly broken down by sunlight, so do not persist in the environment. The use in herbaria of an insecticidal powder, '*Drione*', based on silica dust and pyrethrins is described by Schofield & Crisafulli 1980.

– *DDT* is still available in some countries and has in the past been used on infested specimens. It is extremely persistent and can accumulate to toxic levels in the human body. DDT powder can obscure botanical detail on the specimens and is very difficult to remove. Its use is *strongly discouraged.*

– *Slow release strips* slowly give off a vapour which will pervade the surrounding air space for up to 6 months. They are not suitable for conditions where air change is frequent (e.g. air-conditioning). *Dichlorvos* (*DDVP* or *Vapona*) is the most commonly used. The concentration must be sufficient to kill the pest, but if locally too high, drops of liquid may be formed which can damage specimens and corrode metal (e.g. door hinges etc.). Dichlorvos is reported to have some ill effects on staff.

– *Silverfish bait,* a paste containing the poison *barium fluorosilicate* (or the less effective *sodium fluorosilicate*) has been used in herbaria to control silverfish. Although full details are given in Hall 1988, its use should be avoided as it is a highly hazardous substance banned in many countries.

Protection against fungal damage. Fungal attack resulting from exposure to high humidity over prolonged periods causes decomposition of the plant tissues, and may cause plant features to be obscured and disfigured making them unsuitable for study and dissection.

Fungal damage is a problem in extremely humid tropical areas or at high cool altitudes where fog occurs. In tropical damp conditions the location of herbaria on upper floors and the provision of de-humidifying equipment, if possible, may help. In cool, foggy zones, keeping the building closed and providing a heat source may be helpful. Naphthalene, PDB and LPCP are believed to have fungistatic properties if used in enclosed places. Thymol is quite effective as a fungicide and has been used on important historical documents (see Hall 1988); it is toxic to humans and its use may be banned in some countries (see DePew 1991: 72–81). Painting specimens with alcohol can help arrest fungal growth.

DECONTAMINATION METHODS

Before any dried material enters a herbarium building it must first be subjected to a decontamination process. This should be carried out in a separate building and is *not* to be confused with the methods already described to prevent further attack. Deep freezing is probably the best general method of decontamination.

– *Fumigation* (see above, *Regular fumigation of herbarium rooms and cabinets*) applied in a sealed chamber for several hours is a frequently used method. The gases are poisonous and must **NOT** be used except by people trained in the correct techniques.

– *Controlled atmospheres* (i.e. high concentrations of the atmospheric gases nitrogen and/or carbon dioxide or argon) could be considered for pest eradication. They can be used in a fumigation chamber or a portable PVC container (or 'bubble') instead of the usual toxic gases. Practically no data are available for the control of herbarium pests. The temperature is important since insects are apparently not killed at temperatures below 20°C; 30–38°C may be needed. The required exposure time is considerably longer than for toxic gases. *Stegobium paniceum* can be killed by exposure to 99.5% nitrogen at 30°C for one week. See Pinniger 1991; Strang 1992: 18.

– *Heating* an average size bundle to 45°C for 2–3 hours is a method that has been used, but it is not totally effective. Some herbaria heat to 80–100°C. Damage will be caused to the paper, glue and specimens. See Skvortsov 1977: 131.

– *Gamma radiation* is increasingly used to sterilise food and medical items and may have potential for pest control in herbaria. So far, little experimental work has been carried out, but some evidence suggests it may decrease the strength of some papers (Butterfield 1987); possible effects on the specimens are not yet known. The main drawbacks are high cost coupled with safety restrictions in some countries. See Stansfield 1989; Pinniger 1990 and 1991.

– *Microwave* ovens can be used for small quantities of material (for example a single bundle of specimens brought by a visitor and needed at short notice). It should be noted, however, that specimens must be thoroughly dry. Metal clips and pins can cause charring of the paper and specimens and should be removed. Suggested exposure times on full power: bundles up to 5 cm. thick – 1 minute; 5–10 cm. thick – 1.5 minutes; 10–15 cm. thick – 2 minutes. It must be remembered that not all ovens work at the same power; a test should first be carried out by placing a cube (0.5cm^3) of paraffin wax with a melting point of 58–60°C in the centre of a bundle of newspapers and noting the time it takes to melt completely. On *no* account should the microwave be left unattended, as vapours from the wax could ignite. The temperature 58–60°C is lethal to all stages of most insect pests. Most microwave ovens do not heat evenly and rarely heat right into the corners; there is therefore, the risk that not all pests will be destroyed.

Microwaving can have deleterious effects on the specimens: the viability of seeds and spores will be greatly reduced; pine cones can open; fruit such as nuts can explode; cell structure in fungi may be affected; surface waxes may be destroyed; slight changes may be made to the surface of trichomes and it is possible that there may be effects on the macromolecular chemistry. In addition there is some evidence that glue on mounted specimens can be weakened. If specimens have been microwaved it is good practice to note this on each specimen. See Florian 1986; Hall, A.V. 1988; Hall, D.W. 1981; Hill 1983; Philbrick 1984; Stansfield 1989.

– *Deep freezing* has been used very successfully in the last decade as a safe and reliable alternative to fumigation and it is this method which was extensively tested at Kew before being adopted in 1979 (see Cowan 1980) for decontamination. Provided there is a reliable electrical power supply, this is probably the best method of decontamination for most herbaria. Specimens do not become noticeably brittle, but concern has been expressed for specimens mounted by overall glueing (Egenberg & Moe 1991; see **9. Mounting herbarium specimens** for further discussion). Seed viability may be lowered. See Crissafulli 1980, Hall 1988, Stansfield 1989.

Decontamination by deep freezing

The following notes are intended to provide some guidance to those who may be planning to adopt this method of decontamination. See also **13. Ancillary collections**, *Wood collection*.

The freezer unit

– Top-opening chest freezers should be used because, when opened, there is less cold-loss than with front-opening cabinet units.

– Ensure that the temperature will fall to at least –18°C; most standard domestic freezers will do this. Specialist units are capable of reaching –30°C, and this is to be preferred. Units with internal fans are the best.

– The dimensions should be such that standard-sized bundles can be packed to give maximum space economy. (The Kew units have a capacity of 0.4 cu.m. (14 cu.ft.) and conveniently just takes stacks of bundles with their long axes comfortably occupying the distance between the front and back of the freezer).

– If necessary the units should be lockable to prevent unauthorised opening during disinfestation.

– If money and space permit, install more than one freezer. This will avoid the need to open a unit for the addition of new material part way through the decontamination programme. Each opening causes temperature fluctuation and moisture intake.

Chill-time

The time for bundles to reach the required –18°C at their centre is known as chill-time. The insulating effect of the paper and various other herbarium materials has been tested, and 17 hours has been established as the average. Very woody material or boxed samples etc. may take longer, depending on size and density. Aim for as short a chill-time as possible at a temperature below –18°C. Chill-time will be reduced if the freezers have an internal fan. Adult cigarette beetles have been reported to survive a very slow decrease in temperature to below –18°C (Hall 1988).

Kill-time

The time needed to kill a pest at or below –18°C is known as the kill-time.

- Temperate herbaria will almost exclusively be concerned with the eradication of the drug-store beetle, *Stegobium paniceum.* Controlled testing has indicated that a 3-hour period at a temperature below –18°C is adequate to kill every stage of its life cycle with the exception of a very few eggs which can be killed by a longer period at the low temperature.

- Experiments with other insect species have recorded the longest kill-time to be 9 hours. Hibernation is not generally considered to be a hazard in the deep-freezing procedure. Eggs of some species may survive –18°C (Florian 1986), although temperatures down to –30°C are effective in killing all stages of major pest species. If the freezers cannot reach temperatures as low as –30°C, a double freezing regime can be operated. After the initial freezing at –18°C the specimens are returned to room temperature for a few days before refreezing to –18°C (Pinniger 1991).

Operating time

Combining chill- and kill-times and taking into account practical operating hours, a 48-hour total period sealed in the freezer is recommended as the operating time. This figure is regarded as giving a wide safety margin using freezers which are permanently maintained at the required low temperature.

Maintenance

- Ice build-up in well maintained freezing units should be minimal, but defrosting at the manufacturer's approved intervals is necessary to avoid loss of efficiency of the cushion lid-seal.

- Keep the freezer(s) in as dry an atmosphere as possible to avoid rusting of the casing and to reduce moisture intake when the lid is opened. Very little ice forms in the units if loading and unloading is rapid and the 48-hour decontamination period is uninterrupted. However, ice build-up may be expected to be greater in humid tropical conditions.

- Most units maintain their low temperature for some hours (up to 6–7 hours) after breakdown if they are kept sealed.

- To prevent losing pieces from the specimens and to ensure that only the small percentage of moisture drawn from the herbarium material is reabsorbed by the cold specimens after decontamination, bundles should always be enclosed in sealed polythene bags. Ideally the air should be sucked from the bag with a vacuum pump, but if this is not possible silica gel can be placed inside the bag if condensation seems likely. In the event of unpredicted power failure during unmanned periods, the bags would also help prevent absorption of meltwater.

- After removal from the freezers, material may be taken into the herbarium building where the polythene covering is removed and returned for re-use. If necessary, the specimens can be handled immediately and are certainly ready for general usage after a few hours.

- If the specimens are damp (e.g. they have been left in the rain) they should be dried before they are decontaminated by the deep freezing method; this prevents extra moisture being taken into the units.

- If the specimens have not been thoroughly dried in the press (as frequently happens with succulents) they may turn black as a result of freezing.

REFERENCES

Anon. (1939)
Butterfield (1987)
CCI: N3/1 (1990)
Clark (1988)
Cowan (1980)
Crissafulli (1980)
Croat (1978)
DePew: 72–88 (1991)
Edwards, Bell & King (1980)
Egenberg & Moe (1991)
Eusebio & Stern (1964)
Florian (1986)
Fosberg & Sachet: 65–68 (1965)
Guillarmod (1976)
Hall, A.V. (1988)
Hall, D.W. (1981)
Hill (1983)
Lundell & Kirkham (1966)
Mori et al. (1989)
Philbrick (1984)
Pinniger (1990)
— (1991)
Retief & Nicholas (1988)
Schofield & Crissafulli (1980)
Skvortsov (1977)
Stansfield (1989)
Strang (1992)
Whitmore & Fosberg (1965)
Womersley: 53–54, 67–71 (1981)
Zycherman & Schrock (1988)

5. MATERIALS

Materials for use in herbaria such as paper, board and glue, must be carefully chosen. Although economy can be borne in mind for other items, mounting materials should ideally be of archival quality, or at least the best the institution can afford.

It is important that stocks are always maintained and orders placed well in advance of any items running out. Orders should state precisely what is required, using technical terms if needed, (see below, *Paper Glossary*). Wherever possible samples should be obtained before an order is submitted.

Many materials can be re-used, and these should be returned to the stores for that purpose.

STRING AND WEBBING STRAPS

Purpose: to secure bundles for transit and storage within the herbarium.

Specifications:

- *String* should be strong and of medium weight, preferably smooth. If tied properly the bundle will be secured on all four sides and will keep the correct tension.

- *Webbing straps* – two straps with buckles will be needed for each bundle. Place the straps around the width of the bundle, taking care not to over tighten the first strap, thus distorting the bundle into a wedge-shape. See also p. 213.

Alternatives and economy: webbing straps are expensive but providing they are well made they should have a long life. If the 'herbarium knot' is used, string will last indefinitely since it will not need cutting to release the bundle. The knot can be quickly tied and untied and is therefore economical of staff time. (See fig. 3).

CARDBOARD AND MILLBOARD SUPPORTS

Purpose: placed at top and bottom of bundles of unmounted and mounted specimens to support and protect them during various stages of herbarium procedure and also within packaging for dispatch. Corrugated boards can also be used to aid ventilation in plant presses.

Specifications: the size should be a little larger than the mounting sheet.

- *Cardboard* (corrugated) – Light and ideal for packing loans, saving postage costs. The corrugations should run lengthways to give the best structural support to the card. If widthways, the card tends to fold along the

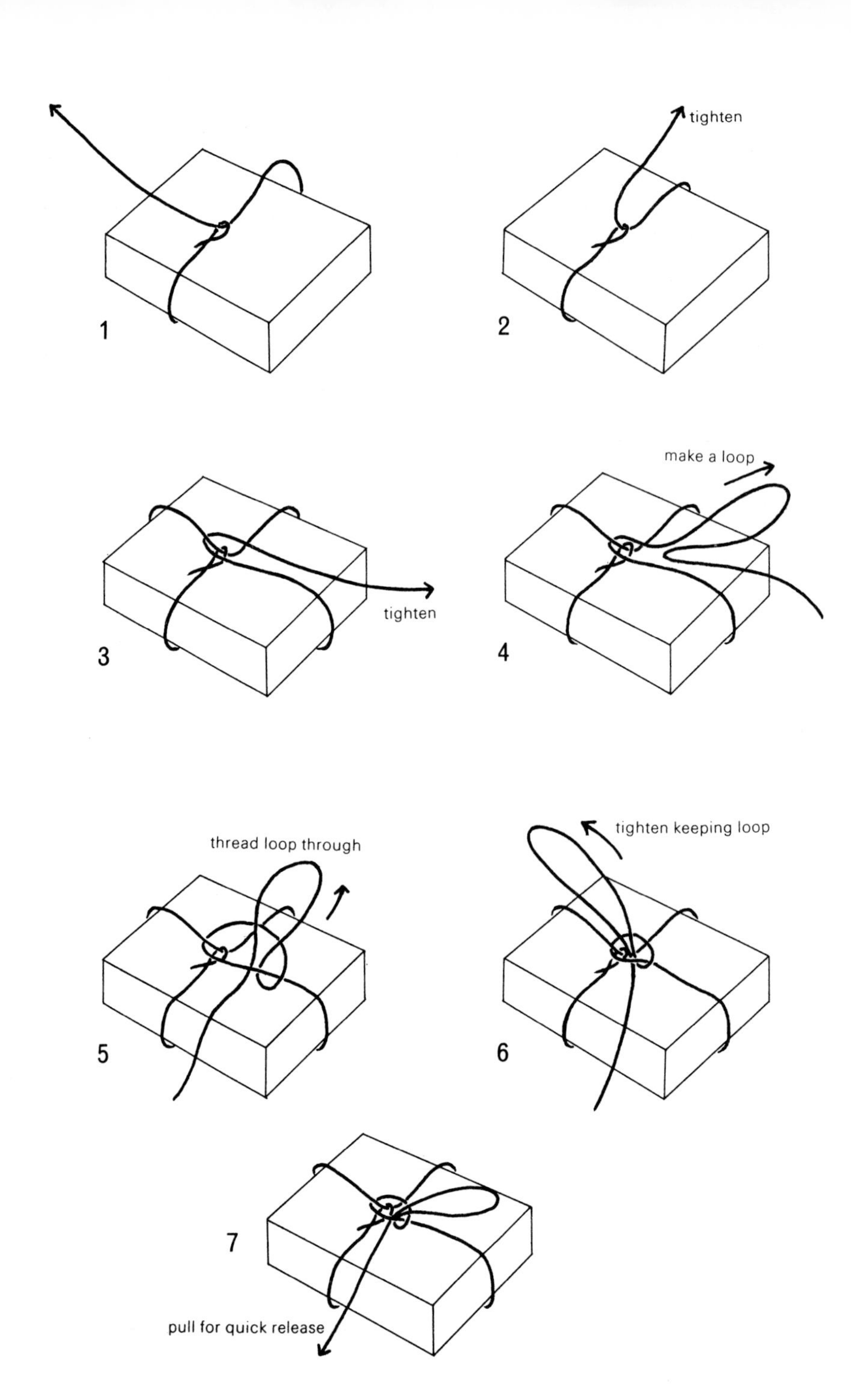

FIG. 3 — The herbarium knot.

corrugations when under tension, and can damage the specimens. Laminated (2- or 3-ply) cardboard will overcome this problem but is somewhat more expensive. Lightweight board will bend and can be cut at edges by tight string. However, this can be prevented by placing card 'pads' folded in a right angle under string at the edges. If required as ventilators the corrugations should run widthways. New stocks should be decontaminated before they enter the herbarium (see **4. Pests and treatments**).

— *Millboard* – dense, strong and rigid (except for the lightest weights), usually grey or blackish in colour. Gives good support to the specimens and will not cut at edges when tied with string. Tends to be heavy and not recommended for dispatch as this adds to the postage costs – may still be worth considering for valuable or delicate artwork or important specimens (especially if only few in number).

HANGING LABELS

Purpose: to identify bundles and display information as to stage reached or next procedure required.

Specifications: strips of thin card or stiff paper c. 7 × 23cm., folded in half with the top glued to the upper board of a bundle so that the lower part hangs down to display the information.

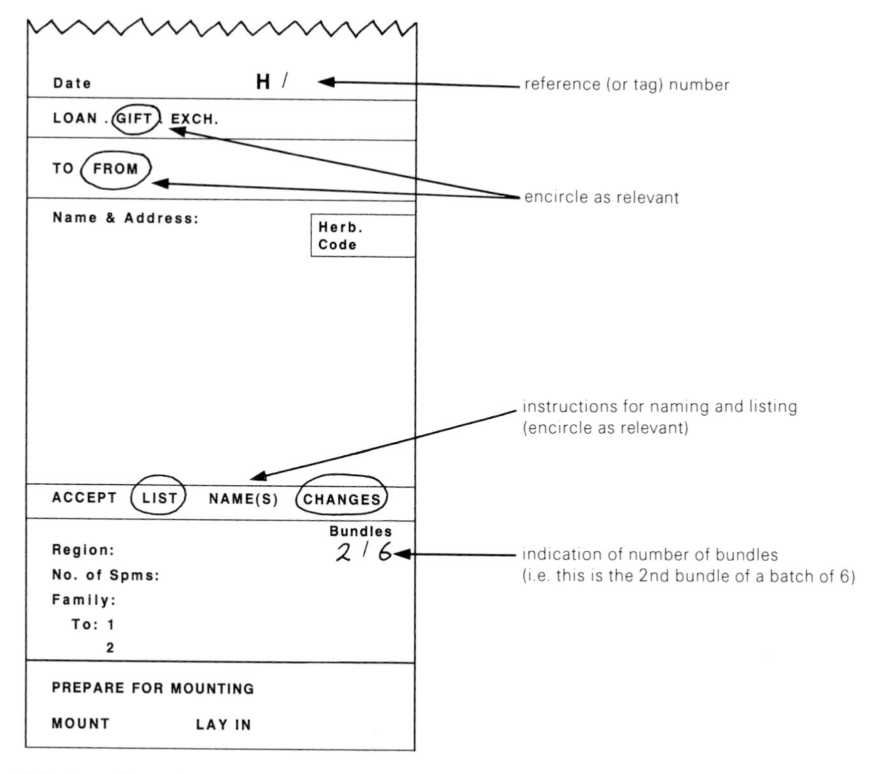

FIG. 4 — Example of a printed multipurpose hanging label.

The information can be printed as a proforma as this standardizes the format and saves staff time. (See fig. 4). Hanging labels can be colour coded, e.g. use a different colour for presented material, loans from your herbarium, loans to your herbarium and duplicates for presentation to another herbarium.

Alternatives and economy:

- Cut up old species or genus covers.
- Save any off-cuts if you trim your own papers to size.
- Request off-cuts from your supplier.

DOUBLE FLIMSIES

Purpose:

- Use to enclose unmounted specimens.
- Use to protect mounted sheets being sent on loan.
- Use in plant presses to enclose specimens so that they need not be handled when the drying papers are changed.

Specifications: very thin but tough paper folders, the same size as a mounting sheet.

Alternatives and economy: can almost be considered a luxury item, as newspapers can be used instead. However, avoid long-term storage in newspaper and dipping in poison baths that may include a solvent of the ink.

NEWSPAPER FOLDERS

Purpose: use to enclose unmounted specimens.

Specifications: cut to size if necessary. Try not to use very new papers as the ink will still be smudgy. Newspaper contains a lot of wood pulp and becomes acidic and brittle with age; it is therefore not recommended for long-term storage of specimens.

Alternatives and economy: newspaper can be used instead of drying paper if that is not available.

MOUNTING PAPER AND BOARDS

Purpose: to support the specimen as a permanent record in the herbarium.

Specifications: use the best white or cream cartridge-type paper available.

- *100% rag* is recommended but chemical wood pulp is an adequate substitute. It should also be buffered to prevent acid migration and be sulphur-free. Mechanical wood or ground wood pulp is not recommended as it becomes acid, discolours and deteriorates. (See below, *Paper Glossary*, 'woodfree').

- The grain should run parallel to the length of the sheet to ensure ease of handling when mounted.

- A *matt* surface, known as 'NOT', allows glues to stick better. (See below, *Paper Glossary)*. A moderately textured surface is preferable to a smooth one.

- *Thickness* (or weight) (See below, *Paper Glossary)*. If possible examine samples before choosing. As a simple test (for minimum weight) it should be possible to hold a sheet of paper horizontally at one end without it bending. As many as four different weights are used at some herbaria, but one relatively heavy one should be adequate. In addition a supply of rigid, white, acid-free boards will be needed for especially robust specimens.

- It is useful to have the institution's name or abbreviation printed in the top left hand corner of each mounting sheet.

- *Size* – the following sizes are most frequently used: 42 × 29 cm. (16½ × 11½ ins.) mostly in USA; 42 × 26 cm. (16½ × 10½ ins.) in Kew and certain other European Herbaria; some herbaria use larger sizes e.g., Natural History Museum 44 × 29 cm. (17½ × 11½ ins.). For economic reasons consider A3 size if you are starting a new herbarium, but the most important governing factor will be the cabinet size. A few extra large specimens may require 'palm-sized' sheets, normally twice the standard size.

Alternatives and economy:

- Good quality is vital; inferior substitutes are not recommended.

- Poor quality paper will deteriorate and remounting will become necessary; the cost in extra staff time would be far greater than the initial cost of better material.

WAXED PAPER

Purpose:

- Sheets placed on top of newly mounted specimens prevents glue sticking to the mounting sheets above while drying under pressure.

- Small pieces can be used to prevent overlapping labels sticking to specimens. The paper is inserted behind the overlapping part of the label while the glue is drying.

- Also used for the field collection of fungi. For this purpose it may be purchased in roll-form.

Specifications: should be slightly larger than the mounting sheet. Strong paper will last longer and can be used repeatedly.

Alternatives and economy: avoid very thin paper.

TRANSLUCENT PAPER

Purpose: to provide a protective flap or 'window' over delicate flowers on mounted specimens. Larger sheets can be used as flaps to protect mounted photographs from scratching.

Specification: must be semi-transparent and must be able to withstand being folded back repeatedly and still lie flat. Needs to be strong as it is intended to protect delicate petals, therefore it should not tear easily. A good quality acid-free paper will fulfil these requirements for a longer period than a poor quality paper. If these 'windows' become damaged, they are very easily replaced.

Alternatives & economy:

- Opaque acid-free paper can be useful but not being able to see the structure covered is inconvenient.
- Polythene tends to crease up, thus no longer protecting the flower. It may also develop static electricity and cling to delicate corollas.

SINGLE FLIMSIES

Purpose: to use as layers to separate specimens arranged (or layed-out) in piles ready for mounting.

Specifications: as double flimsies but a single sheet.

Alternatives and economy: use not essential, especially when specimens are layed-out and mounted in a single process.

SPECIES COVERS

Purpose: to group and protect specimens of the same species in the herbarium. Species covers containing specimens are ideally placed inside a genus cover. (See below, *Genus covers*).

Specifications: use acid-free, light weight but tough paper with the grain running parallel with the fold, (see below, *Paper Glossary*). Species covers should be slightly larger than the mounting sheet when folded in half.

Alternatives and economy: inferior grades of paper can be substituted providing they are acid-free.

TYPE COVERS

Purpose: to protect type specimens, retain any fragments that may fall off them, and to draw attention to the specimen.

Specifications:

- Acid-free strong paper slightly larger than amounting sheet when folded with overlaps at all four edges. The top and bottom overlaps need only be c. 3 cm. wide; the left-hand overlap should entirely cover the specimen, while the right-hand (outer) overlap could be 10 cm. or more wide to produce a half-flap which can bear the relevant information (see fig. 25).

- Type covers should either contrast in colour with genus and species covers or have the top and bottom edges printed with bands of a bright colour (usually red), with the word 'TYPE' and name of the herbarium also printed.

Alternatives and economy: keep type specimens in separate species covers, clearly marked 'TYPE'.

GENUS COVERS

Purpose: to group specimens of one genus and protect them in the herbarium. Specimens are usually subgrouped into species covers before they are placed in the genus cover. See Jain & Rao 1977: 69–71, fig. 27.

Specifications:

- Acid-free thin card or thick paper cut with grain running parallel with the fold. The size should be such that when folded with a spine (two folds c. 2 cm. apart) it is slightly larger than the mounting sheet. An alternative design has a second spine so that the specimens are protected on both sides and a flap is formed. This type of genus cover also limits the number of specimens contained in it (i.e. is difficult to over-fill). Genus covers may be pre-scored and pre-folded but more usually are supplied flat. Use a ruler and sharp knife to score (i.e. partially cut through) the lines of the folds, then fold using a suitable implement, e.g. bone folder or ruler. The better the quality of card the easier it will be to fold.

- Genus covers are usually buff or another neutral colour, but some herbaria use different coloured genus covers to indicate geographical areas.

Alternatives and economy:

- Can be dispensed with in small herbaria providing the species covers are robust.

- If a genus cover bears incorrect information re-use by turning back-to-front or upside down.

A Small capsules made of white bond paper

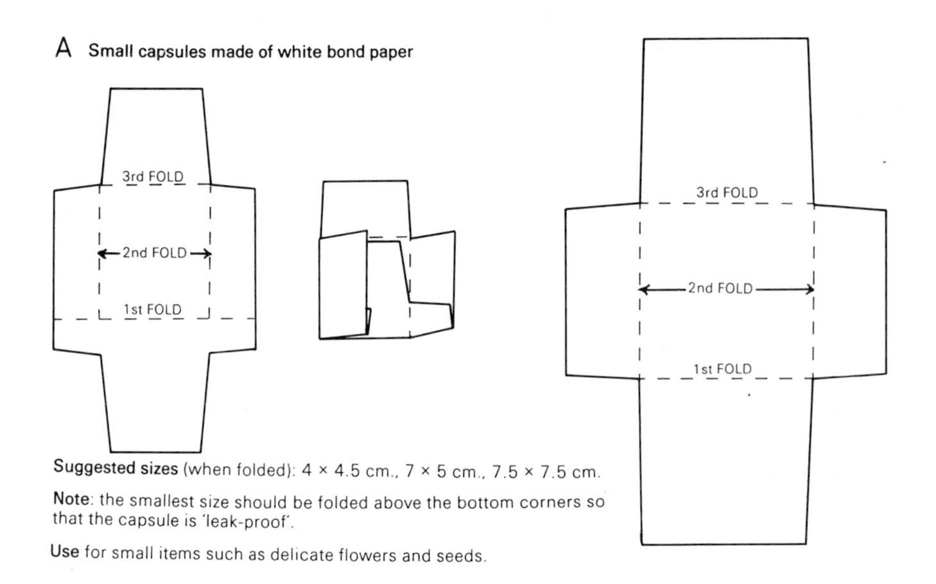

Suggested sizes (when folded): 4 × 4.5 cm., 7 × 5 cm., 7.5 × 7.5 cm.

Note: the smallest size should be folded above the bottom corners so that the capsule is 'leak-proof'.

Use for small items such as delicate flowers and seeds.

B Large capsules of strong manilla paper

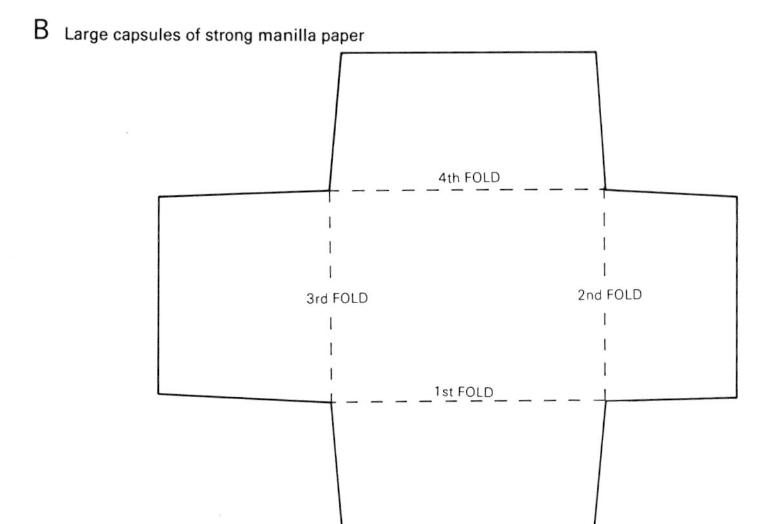

Suggested sizes (when folded): 20 × 11 cm., 17.5 × 12.5 cm., 12 × 9 cm.

Note: the side flaps should allow a good overlap when folded. Preferably one side flap should be shorter than the other so that the overlap does not fall directly in the centre (corresponding to the thickest part of the filled capsule).

Use: for detached leaves and large items. If there are also small delicate flowers or seeds, put them inside a smaller capsule and place inside the large one.

FIG. 5 — Suggested templates and folding sequences for ready-cut capsules.

PAPER CAPSULES (ENVELOPES)

Purpose: glued to mounting sheets to contain and protect small portions of the specimen, or very small specimens.

Specifications:

- Acid-free white or buff-coloured paper.

- There is a variety of sizes and styles, but when folded and glued to the sheet the capsule should open flat, and stay closed without the aid of a paper-clip (unless it contains a bulky portion), and not allow small seeds to escape from the bottom corners. Large herbaria order stocks of ready-cut paper capsules in different sizes and weights of paper, but care must be taken to fold them in the correct way. See fig. 5 for suggested templates and folding sequences.

Alternatives and economy: perfectly adequate capsules can be folded from a rectangular piece of paper (such as a sheet from a writing pad). See fig. 6 for folding sequences.

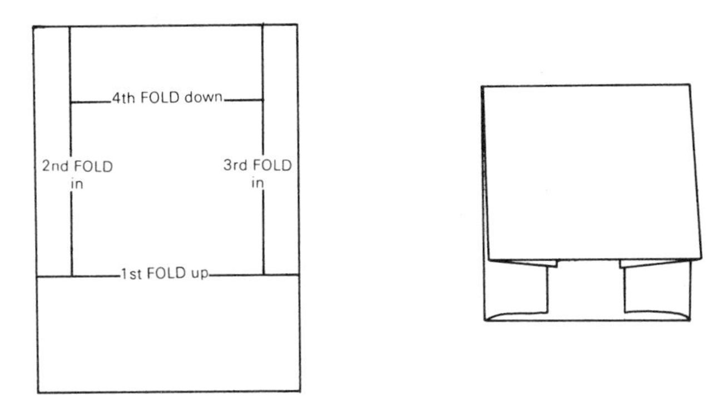

FIG. 6 — Capsule folded from a rectangle of paper.

Not recommended:

- Conventional envelopes; these do not open flat and delicate material can be damaged when removed for study.

- Capsules made from a rectangle of paper by folding the sides underneath. The capsule will have to be curved up at the sides when opened causing damage to delicate contents.

- Capsules made from brown kraft paper, such as used for wrapping paper. This is acidic; the capsules will crack along the folds, especially if left exposed to the sun.

OPEN-TOPPED ENVELOPES

Purpose: used to contain fungi specimens or photographic prints of herbarium specimens for incorporation in the herbarium collections. The envelope will form a pocket when glued to the sheet.

Specification: Acid-free, wove white or buff-coloured, stiff paper.

Not recommended: do not use for portions of herbarium specimens, these should be put in paper capsules, see above.

MOSS PACKETS

Purpose: to contain bryophyte or fungal specimens, either for mounting on herbarium sheets or loose-filing.

Specification: strong wove, acid-free paper such as cartridge paper. The design is like a conventional envelope but the flap should not be gummed. Recommended sizes: 12 × 9 cm., 15 × 11.5 cm., and 18 × 12.5 cm.

Not recommended: do not use for portions of herbarium specimens, which should be put in paper capsules, see above.

TRANSLUCENT PACKETS

Purpose:

- To contain small seeds or fungi inside a paper capsule.
- To contain small items in transit, e.g. buds for pollen samples or seeds.

Specification: small sealable envelopes made of thin translucent paper or film. Polyester or glassine may be used, see below, *Paper Glossary*.

Alternatives: polythene packets can be substituted for many purposes. The self-closable or 'zipper' type is ideal.

Not recommended:

- Delicate flowers should not be placed inside glassine packets; it is difficult to slide the flowers out without damaging them.
- Avoid cellophane packets as these have a short life.
- Polythene packets should not be used for microscopic items which will cling by static electricity.

TRANSPARENT SLEEVES

Purpose: to protect old or delicate herbarium sheets or e.g. grass specimens with loose inflorescences etc.

Specifications: Polyester (not cellophane or polythene); sleeves should be slightly larger than the sheet and open-ended by preference. Only one side need be clear.

Alternatives: protect sheets with type covers, clearly noting if 'non-type' material or place in individual double flimsies or species covers.

PAPER-CLIPS

Purpose: to hold the edges of paper capsules together when the contents are lumpy.

Specifications:

- Continual-loop plastic clips are best for humid tropical conditions. They are also the cheapest. Brass paper-clips are also available and do not rust and stain the paper.
- 'No-tear' metal clips are acceptable for drier conditions where they are unlikely to become rusty.

Not recommended: ordinary office clips tend to rust, and to catch the edges of the sheet above and can cause tearing. (See fig. 7).

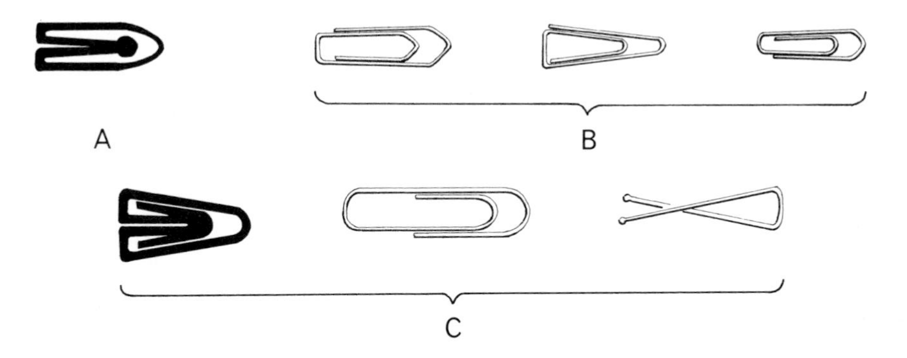

FIG. 7 — Types of paper-clip: **A** continual loop plastic clip; **B** no-tear metal clips; **C** office clips, *not* recommended for use on herbarium specimens.

WRITING INKS AND PENCILS

Purpose: for use on labels, genus covers, species covers and type covers.

Specifications:

- Ink must be permanent and preferably waterproof or water-resistant. Choose black (or blue-black), other colours should be avoided. Use Indian ink for dip pens and choose brands such as 'Rapidograph' or 'Pelican' inks specially designed for fountain or stylus pens. It is now possible to buy disposable pens with water-resistant ink, in a range of nib sizes. Disposable pens with spirit-resistant inks are also available; they should be used for writing labels to be placed inside spirit bottles, but first test the ink in spirit to make sure. Some inks (especially coloured inks) will fade if exposed to light. Black waterproof inks will usually be light-resistant since they contain lampblack or other carbon pigment (sometimes reinforced with other colouring material).

- Ink for stamp pads should also be of permanent quality.

- Pencil – use a medium soft lead such as HB, 2B or F. Too hard a lead will dent the paper and be difficult to erase, while too soft a lead will smudge.

ADHESIVES FOR MOUNTING

Purpose: mounting specimens, attaching labels, mounting photographs and paper capsules, etc.

Specifications: Ideally the adhesive used must be neutral or slightly alkaline, extremely long-lasting and reversible (i.e. water soluble).

The best types are:

- *Starch Paste.* This is very economical and can be made from either wheat-starch or rice-starch (10% weight/volume). The mixture must be cooked until it is gelatinous. A few drops of a fungicide solution can be added with caution; thymol, however, is toxic to humans and banned in many countries.

- *Polyvinyl Acetate (PVA).* The *copolymer* type is the one to use as this is internally plasticized, designed for permanence and remains flexible. The adhesive is a white emulsion which becomes transparent and glossy when dry; it dries relatively quickly. It may be thinned with water to the required consistency. In general it tends to buckle the paper less than starch paste. There are many formulations and it is now possible to purchase special PVA glues of archival quality which are neutral and more readily water-soluble than previous formulations. If in doubt check the formulation; the manufacturers may have modified it or introduced a new formula. Woodworking PVA adhesives (e.g. 'Evostick Resin W') have in the past been widely used. Although they are easily obtainable, economic and good general purpose glues, tests have shown them to become somewhat brittle and less water-soluble with age.

- *Methyl cellulose.* This adhesive has been recommended for archival work for many years. It is supplied in powder form and must be mixed with water, but unlike starch paste it need not be cooked. It is less strongly adhesive than PVA glues, dries more slowly, is pest resistant and is easily reversible with cold water. Methyl cellulose may be used in combined formulations with starch pastes or with PVA adhesive, also magnesium carbonate powder can be added as a protective alkaline reserve. It is the usual adhesive on modern gummed labels.

- *Sodium carboxymethyl cellulose* has very similar properties to methyl cellulose and can be used as a substitute. It may, however, be somewhat less archivally stable.

Other adhesives that have been frequently used in herbaria include:

- *Latex adhesives.* These have been used for general mounting and for the specialist mounting of delicate aquatics (see **9. Mounting herbarium specimens**). The main constituent is polyisoprene (natural rubber) and the usual solvent chloroethane; however, water-based formulations are also available. Only sufficient for 2–3 days should be diluted at a time. The advantage of latex glues is that buckling of the paper due to the absorption of moisture is avoided. However, it is not of proven archival quality and tends to become insoluble with age.

- *Archer's plastic adhesive.* A thick transparent adhesive used for 'strapping' or 'blobbing' specimens by the Archer method (see **9. Mounting herbarium specimens**, *The Archer method).* Archer's plastic adhesive varies with the formula used but it is based on ethyl cellulose (ethocel) and Dow resin dissolved in toluene (or tolulol) and methanol, or the English equivalent 'Bexol SR 101' with a toluene solvent. Various formulations are discussed by Croat (1978).

  The compound can become 'tacky' in warm humid conditions and it may be adversely affected by the insect repellant paradichlorobenzene (see **4. Pests and treatments**). The adhesive dries quickly, making errors difficult to clean up; it is also non-reversible. The toluene fumes present a health hazard and this adhesive should therefore be used in a fume-cupboard.

- *Gelatine/glue.* Glue was traditionally made from animal skins, bones etc. and is an impure product. The term gelatine is used for a more pure product from the same sources, but glues and gelatines intergrade. *Fish glue* is an impure gelatine prepared from fish heads, bones and skins; a glue made from the skins alone is purer. *Isinglass* is a very pure fish gelatine made only from swim-bladders.

  Gelatines and glues tend to be acidic. They are liable to bacterial attack and in time dry and may become ineffective; they are, however, water-soluble. Glues were widely used in herbaria in the 19th century, but now the specimens sometimes need repair.

– *Gums* (in the strict sense) are of vegetable origin: *gum arabic* is obtained from *Acacia* spp. and *gum tragacanth* from *Astragalus gummifer.* Gums have rather similar properties to glues and gelatines.

  See CCI 1990; Clark 1986 & 1988; Croat 1978; Digby 1985.

Problems: some glues weaken with age or lose their adhesive qualities completely, or they may become brittle or soften and become sticky. Some cause staining or can alter the colour of specimens, especially of petals. Many commercially available adhesives are acidic and will do this. A neutral adhesive should not stain but it may be more susceptible to mould growth. The following are recommended *only* for short term office use:

– *Solid polymer resin* adhesives (e.g. 'PrittStick'). Quick and convenient but not likely to endure.

– *Rubber cement* (e.g. 'Copydex' and 'Cow gum'). These adhesives oxidize, lose their bonding ability and stain.

Recommendation: Place small, sealable containers of the approved adhesive with applicator in convenient locations in the herbarium for the use of scientific staff and visitors who may wish to attach new determination slips or paper capsules to the specimens. Ask visitors and staff not to use convenience products such as 'PrittStick' on herbarium specimens.

TAPES

Purpose: mounting specimens:

– By strapping the specimen to sheets.

– For covering stitches on reverse side of mounted sheets.

– For mounting floral dissections (see **24. The dissection of floral organs and preservation of the results**).

Specifications: the adhesive for *gummed tapes* should be pH neutral and easily reversible, i.e. soluble in water; usually methyl cellulose is used. The fabric may be either:

– Linen tape as used by bookbinders. The tape should be cut into strips which have the dominant threads (the warp) running lengthwise for maximum strength.

– White (or brown) paper is adequate for protecting knots; it should be acid-free and as strong for its weight as possible.

Self-adhesive, transparent, polyester tapes are obtainable and said to be of archival quality. The adhesive is acrylic.

Alternatives and economy: as an alternative type of strapping use linen thread but remember to cover stitches on the reverse side of the mounting paper with gummed paper, or use the 'Archer method' (with plastic adhesive) see **9. Mounting.**

Not recommended:

- Avoid self-adhesive cellulose tape, e.g. 'Sellotape' or 'Scotchtape' and similar products, except for short term office use. In time the clear tape becomes brittle and falls off, while the adhesive often spreads beyond the edges of the tape and may strongly stain the mounting paper.

- Avoid surgical tape (plasters), as the adhesive spreads out and collects dirt, becoming messy.

- Strips of non-archival paper should be avoided as they will become brittle and discoloured with age.

Caution:

- Gummed tape must be kept dry before use. A humid atmosphere will cause it to stick to itself or to packaging.

- Cellulose or plastic tapes deteriorate with age and can be decomposed by some fumigating agents.

THREAD

Purpose: to attach entire or parts of specimens to mounting sheet, especially to reinforce the mounting of bulky parts of specimens such as twigs and larger fruit.

Specifications: strong unbleached linen thread with a satin finish, such as used by bookbinders, is recommended. Several thicknesses are available; strength increases with thickness, but one should be chosen which is not too thick for ease of use. Threads are measured in sizes, e.g. 12–3, 16–4, 30–2 etc., the higher the size number the thinner the thread.

Alternatives and economy: cotton thread is used by some herbaria for general mounting. However, it breaks easily and must be tied off repeatedly. It is not ideal for reinforcement as it may deteriorate badly with age. Coating cotton thread with beeswax helps to strengthen and protect it.

PAPER GLOSSARY

Terms used in the paper industry can be confusing and it is important to understand them when ordering supplies and using materials.

Acid-free – paper that is free from acid content (i.e. tending towards alkaline) or from other substances likely to have a detrimental effect on the paper.
A term that is used very freely, and which may not in all cases indicate a product with good permanent properties. Although a paper may be acid-free at the time of manufacture it can become acidic due to acid formation from residual chemicals left by bleaching or sizing, from the absorption of atmospheric pollutants or from *acid-migration* from adjacent poor quality papers.

Buffering Agent or Alkaline Reserve – calcium or magnesium compounds added to the paper pulp that help to neutralize acids within the paper and from the environment. Usually desirable, but not recommended for use with some items, e.g. 19th century photographs.

Cold-pressed – see NOT.

Foxing – a widely used term to describe the rusty brown spots or stains that can develop in paper, particularly if of poor quality. Foxing develops in response to high humidity and can be caused, in part, by impurities in the paper. It can migrate to adjacent sheets of paper.

Furnish – the ingredients put together to produce a specific type of paper, includes fibres, fillers (e.g. kaolin), colouring agents, size etc.

Glassine – thin, fairly tough, glazed and nearly transparent paper that is used widely for storage of photographic negatives, and can be written on. It can, however, become brittle and eventually crumble; also at high temperatures and high relative humidity it can stick to itself or to items contained in it.
Transparent and translucent papers may be made from poor quality materials are often treated extensively with chemicals such as sulphuric acid, ethylene glycol, etc. to produce transparency.
Unless the specification is known, such papers should be avoided where permanence is a requirement.

Grain direction or Machine direction – the alignment of fibres in a sheet of paper caused by the flow of the wet paper on the papermaking machine. (See fig. 8). The paper will tear and fold more easily in the direction of the grain. It also tends to expand and contract in this direction as moisture is gained or lost.

Grammage (*G/m² or gsm.*) – the weight of paper and board expressed in metric terms.

Handmade paper is equally strong in all directions because the fibres are meshed at random; it is however, prohibitively expensive.

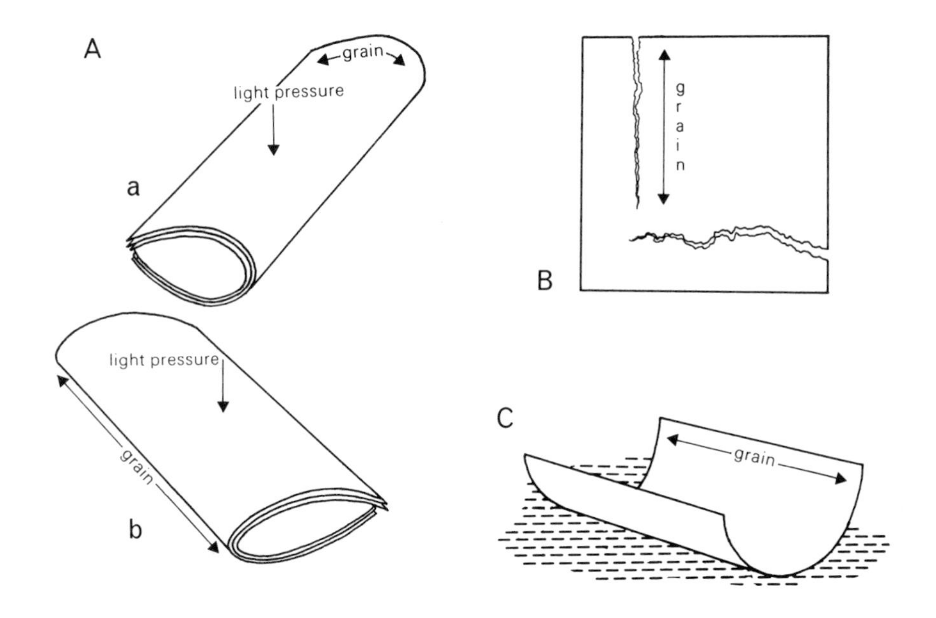

FIG. 8 — Methods of testing grain direction in paper. **A** take a few sheets of paper and lightly bend across the length and across the width – more resistance will be felt when the grain runs at right-angles to the fold (i.e. in **a**); **B** tear the corner of a piece of paper in both directions – the tear made in the direction of the grain will be straighter and more easily made; **C** place a small piece of paper on a dampened surface – the paper will curl at right angles to the direction of the grain.

H.P. – (*Hot-pressed*) – the paper has a smooth surface achieved by passing it through hot metal plates or rollers during papermaking. (See *NOT*).

Laid – a paper which shows the marks of the wires in the mesh on which it was made. On holding it up to light a translucent striped pattern can be seen. These are the laid lines; the widely spaced vertical lines are known as the chain lines.

lb. – the weight of 500 sheets (a ream) of paper of a given size. An imperial measurement (pound) becoming obsolescent.

Lignin – a component of the cell-walls of woody plants. It is an undesirable impurity in paper.

Lignin-free – see wood-free.

Machine-made paper – paper made by forming the pulp on a wire mesh. The fibres align themselves to the direction of the mesh giving the paper a *machine-direction* or grain (see *Grain direction*). Cartridge paper is a machine-made paper.

Mechanical wood or groundwood pulp – produced by mechanical action only in the presence of water. Thus only water-soluble constituents of the wood are lost in the process. Lignin remains in the pulp and contributes to the acidity, brittleness and discoloration of many poor quality modern papers.
Most newspapers are made from groundwood pulp.

mil – unit of thickness (of paper or board) equal to one thousandth of an inch (0.001").

Mould-made paper – paper formed on a cylinder mould. Because the fibres tend to be meshed at random it is more stable in all directions than machine-made paper, but less than handmade paper. It is likely to be too expensive for herbarium use.

NOT (i.e. *Not* hot pressed) – the slightly rough, unglazed surface of a paper. In the USA the term *cold-pressed* is used. See *H.P.*

pH – (potential of hydrogen). A measure of acidity and alkalinity. Pure water has a pH of 7.0 (i.e. neutral). Acid solutions have a pH value lower than 7.0 and alkaline higher.

point – unit of thickness (of paper or board) equal to one thousandth of an inch (0.001″), i.e. 0.050″ = 50 points.

Polyester (or *plastic polyethylene terephthalate*) – a transparent, colourless, tensile and chemically stable film. It comes in different weights or thicknesses, often measured in *mils*. Common tradenames are '*Mylar*' and '*Melinex*'. *Polypropylene* is also chemically inert and quite strong, but it is softer and not so clear as polyester. *PVC* (*polyvinyl chloride*) contains chlorine and plasticizers, and should therefore be avoided.

Rag – traditionally paper was made using linen, hemp and/or cotton rags, as a source of cellulose fibres. In modern usage the term can also refer to paper made from cotton, or cotton linters fibres (i.e. short fibres which stick to the seeds after a first ginning) that have not been previously used in textiles.
Many other fibres have been used for papermaking. These include leaf fibres, e.g. manilla, and grasses, e.g. esparto grass and bamboo, or palms, e.g. rattan.

Rough – paper with an unpressed surface texture, see *HP* and *NOT*.

Sizing – reduces the absorbency of paper so that ink will not bleed. A synthetic size may be included in the *furnish* (internal sizing) or a coating of starch or gelatine may be added to the surface of the sheet (surface sizing); surface sized papers are usually internally sized. Because many sized papers tend to become acidic they are usually *buffered*. Drying papers and blotting papers are usually unsized.

Substance – the weight of paper expressed in lb. (pounds) per ream or G/m^2, (see *grammage* and *lb.*). Heavyweight paper is thick and lightweight paper is thin. In the USA the thickness of the paper may be specified (see, *mil* and *point*).

Weight – see *substance.*

Woodfree – this is one of the most confusing terms and also one of the most important to understand.
'Woodfree' = Chemical Wood Pulp: made either from softwoods or hardwoods and treated chemically using the sulphite, modified sulphite, sulphate or soda processes. The chemical treatment dissolves most of the ligneous material. Pulps may then be bleached to produce a better colour and a more pure product. N.B. 'Woodfree' does *not* denote rag paper.
Chemical wood papers can be produced to high standards with excellent durability properties e.g. alpha-cellulose papers.

Wove – a paper with an even texture made on a continuous close meshed wire belt. It lacks wire or laid lines (see *laid*).

REFERENCES

CCI N11/4 (1990)
Clark: (1986)
— : (1988)
Croat: 215–217 (1978)
DePew: 3–44 (1991)
DeWolf: 82–90 (1968)
Digby: (1985)
Fosberg & Sachet: 55–59 (1965)
Franks: 30–32 (1965)
Jain & Rao: 69–71 (1977)
Tétreault & Williams (1992)
Womersley: 72–73 (1981)

6. LABEL DESIGN AND PRODUCTION

The label is one of the most important parts of a herbarium specimen. It provides essential information about the specimen such as its name, its collector, and where it was collected. Without such information the specimen will be of little or no use. In the early days of plant collecting labelling was generally poor in quality; labels were sometimes printed, particularly for large collections, but in many cases they were handwritten, often on scraps of paper. Such handwriting is often very difficult to read. (N.B. Burdet (1979) gives many useful examples of important early handwriting). Today much greater attention is paid to labelling, although it can still vary in the type, quality and quantity of information given. In addition most herbaria adopt their own style of labelling, depending on their needs.

Archival considerations

Paper used for labels should be bond paper of good quality, unglazed and acid-free (see **5. Materials**). Paper of the correct quality and format is available for both typewriters and most computer printers (see **29. Introduction to Computers**, *Stationery and supplies*). Gummed labels should be of archival quality. It is now possible to purchase foil-backed, self-adhesive labels of archival quality for use on carpological boxes, microscope slides, etc. Non-archival self-adhesive labels of the type used in shops must not be used; they drop off after a few years and can become transparent; gummed labels could be used or ungummed labels attached with a permanent adhesive. A special strong paper, 'Resistall', has been developed for use in spirit (wet) collections.

Since they are liable to deteriorate with age, correcting fluids such as 'Tipp-Ex' should be avoided on herbarium labels.

Printing labels

Labels may be printed by a commercial printer, in which case the fewer the words printed the cheaper the label, and the larger the quantity ordered the smaller the cost per label. In the long-term a cheaper alternative may be to purchase a small hand-operated printing press. This will allow labels of any type to be printed as and when required. A further method is direct offset printing from paper plates, which depends on the availability of good typing facilities and an offset printing machine (see Womersley 1981). Ideally, details about the specimen should be typewritten on the label, since this provides the greatest clarity. However, if typing facilities are not easily available they may be handwritten in permanent ink. Duplicate labels, when required, can be photocopied; carbon copies are an alternative, but their quality is often poor.

Preparation of labels by computer

High quality, detailed labels may be produced using a personal computer together with a printer. This is particularly useful when labels are being made for specimens of a wide range of species from the same habitat, locality, collector etc., or vice-versa. Production of the labels is fast (an average label printed in c. 5 seconds), and a range of printing styles and sizes can be obtained.

Not all computer inks are of archival quality; a photocopy may be advisable.

Types of label

Basic data labels. These labels provide the essential information about the specimen, and should be present on all specimens. The following items should be included:

- Herbarium name – sometimes given in Latin, particularly by the older European herbaria; the name can also be abbreviated (see fig. 9, E & G). Duplicate specimens sent from one herbarium to another should have *ex* (= from) or *from* in front of the name of the donating herbarium. (see fig. 9, F & H).
- Scientific name.
- Vernacular name(s).
- Collector's name, and collection number if any.
- Date of collection.
- Collection locality, preferably with latitude and longitude co-ordinates.
- Habitat/ecological notes – including altitude, if available.
- Supplementary data – this can include notes on the habit of the plant, flower colour and any other features which may be important for identification but cannot be seen directly from the specimen.
- Uses.

The information given on the label will obviously depend on the quantity and quality of data provided by the collector. However, every specimen should at the very least have collector's name and number, locality and date on the label, otherwise the specimen will be of little or no use.

Data labels are often printed with the name of the herbarium at the top of the label, and with sub-headings for each item of information to be filled in (fig. 9,A). Some labels may only have the name of the herbarium printed on them (fig. 9,B) or with the name of the collector and country (fig. 9,C & D). When many specimens have been collected from one particular locality, it is often useful to have habitat and ecological data printed on a separate label, so that duplicates of these labels can be put with each specimen from that locality (fig. 9,I). This saves time having to write out the same information for each specimen. (See **30. Collecting and preserving of specimens** and figs. 51 & 52).

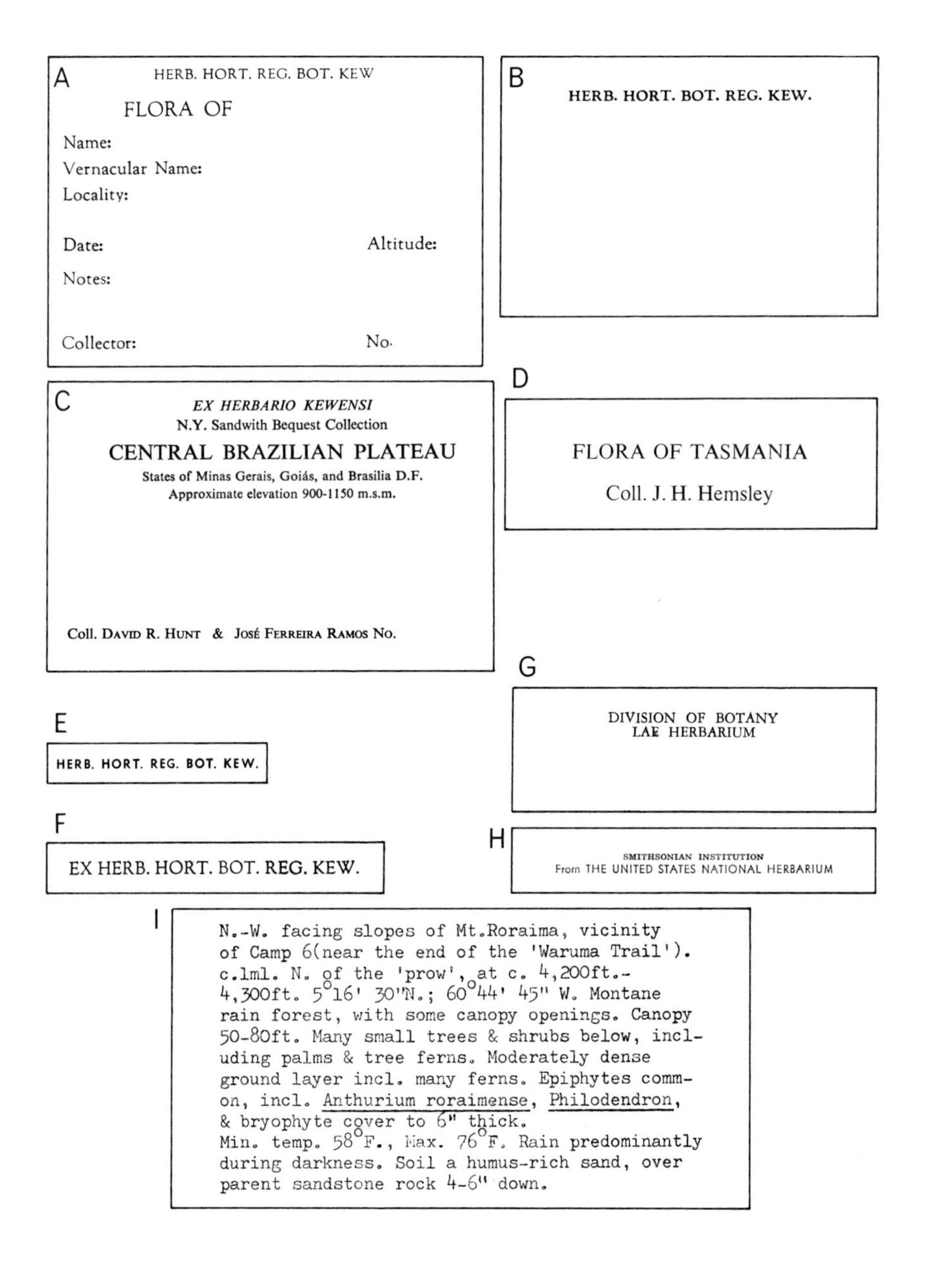

FIG. 9 — Examples of labels. **A–D** basic data labels; **E–H** institution identification labels; **I** supplementary habitat label.

Other types of label. These may be present on the sheets in addition to the basic data labels. Gummed labels are convenient in non-humid conditions.

- Determinavit and confirmavit labels, and labels to indicate that a specimen has been noted in a special research project (e.g. 'Seen for Flora of ...'). (fig. 10).

DET.. 19..............

Seen for "Flora of Egypt"
Vol. , Fam.

Det . . . . 19 . . . . .

DET.. 19..............

Flora of Tropical East Africa

Det. Date

Confirmavit
L.L. Forman 198

Seen for
"Flora Zambesiaca"

Seen for Revised
Edition of F.W.T.A.

HERBARIUM KEWENSE

Drawing prepared for the "Kew Bulletin,"
Vol. , No. ( ).

WAKEHURST SEED BANK VOUCHER
NAME ....
VERIFIED BY ....
DATE ....

FIG. 10 — Examples of determination labels and labels to indicate special studies etc.

— Institutional labels – used if the herbarium sheet or data label does not include the name of the herbarium (fig. 9,E & G).

— Labels indicating that specimens have had small pieces removed for anatomical, palynological or phytochemical study. These may be printed on different coloured paper to indicate the discipline concerned (fig. 11).

Anatomy slide in Jodrell Collection
Published data :..
....

Pollen material from this
specimen sent to

Date:

Sampled for chemical constituents Date :
Investigator : Ref.no.
Compounds found :

ARC Biochemical Laboratory, RBG, Kew.

Pollen material taken from this
specimen for study at the
Royal Botanic Gardens, Kew

Seed removed for S.E.M.
scanning - - - - - - - - -

FIG. 11 — Examples of labels indicating that portions have been removed for special studies.

— Type labels – indicate that the specimen is a type of the name shown on the label. These labels are usually printed in red (fig. 12).

HOLOTYPE	NEOTYPE
of	of .... selected by .... in ....
ISOTYPE	**LECTOTYPE**
of	of .... selected by .... in ....
SYNTYPE	**TYPE**
of	of ....

FIG. 12 — Examples of type labels.

Labels for genus covers. – These give information on family and genus and/or geographical region for the specimens contained within the genus cover. Labels showing the geographical region may be printed on coloured paper, with different colours corresponding to different regions. Local and special herbaria may find labels for genus covers unnecessary (fig. 13).

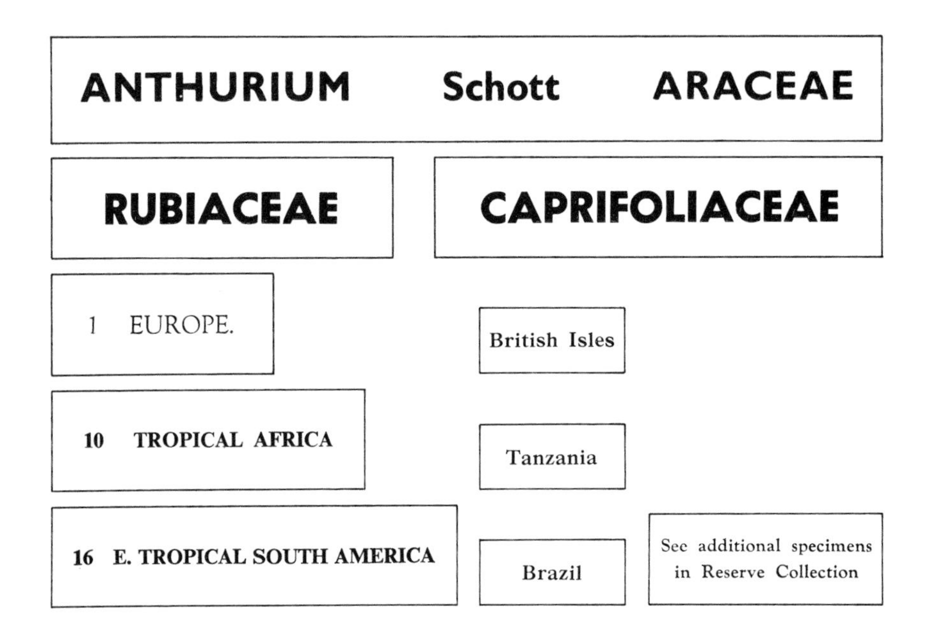

FIG. 13 — Examples of labels for use on genus covers.

REFERENCES

Burdet (1979)
DeWolf: 83–84 (1968)
Fosberg & Sachet: 68–71 (1965)
Franks: 11–13 (1965)
Womersley: 65–66 (1981)

7. CENTRALIZED ACCESSIONING, RECORDING AND DISPATCH PROCEDURES

However small the institution there are advantages to be gained by the centralization and standardization of accessioning, recording and dispatch procedures:

- It frees scientific staff to spend more time on research.
- It provides readily accessible, standardized records of all accessioning, dispatch and loan procedures.
- It provides a facility for the correlation and compilation of determination lists, and for the monitoring of loans to and from other institutions and the dispatch of reminder letters for overdue loans.

RECORD KEEPING AND RETRIEVAL SYSTEMS

Correspondence

All letters should be kept on file and be retrievable; it is important that they are not discarded after the necessary action has been taken in case queries arise in the future. Very old correspondence can be of great historical value, in which case it should be kept under archival conditions. Many institutions have a centralized records department which administers the long-term storage and filing of correspondence.

Recording accessions, loans and duplicates for presentation

In choosing a recording system, consider the following points:

- Does it fit present and forseeable future requirements?
- Is it flexible enough to be altered if the nature or volume of the work changes?
- Can it be simply understood and easily used by the staff who may wish to consult it?
- Is it cost-effective, i.e. can the cost be justified? Are the basic materials or maintenance (e.g. computers) readily available? Are the demands on staff-time heavy?

There are many ways to store the necessary records; some are more convenient and cost-effective than others:

- *Simple ledgers* for recording incoming and outgoing material. These tend to be inflexible, as all entries will be restricted to chronological order. Separate ledgers will be needed for the different functions e.g. accessions, incoming loans, outgoing loans etc. If the ledgers are large some indexing system may be needed to facilitate finding past entries.

– The records can be kept on *loose-leaf sheets* or forms and filed (or placed in folders) under specific headings (e.g. accessions, outgoing loans or under a heading for each individual institution). All paper work must be dealt with promptly and accurately to prevent pages being lost or misfiled.

– If using a *computer* (see **29. Introduction to computers**), remember to keep at least one, preferably 2 or 3 back-up sets of data. All staff needing to use the system will require some training.

– *Tag-system.* A tag number (i.e. reference number) is allocated in strict chronological order to every incoming and outgoing consignment of specimens. The numbers include a reference to the year, e.g. 100/89, 101/89, 102/89 etc., and these are printed on the tags (= forms) together with various headings covering all the information needed relating to the specimens. The tags are bound in a book and each tag is duplicated. The tag itself, i.e. the top copy, is perforated and when removed from the book it is attached to the relevant correspondence; the duplicate page is not perforated and remains in the book. The details of the consignment are entered in the book on the appropriately numbered tag and any notes or instructions are added, e.g. 'loan from ...' or 'received for naming' etc. When the

Date tag closed | Sender's Ref.no./Date

INCOMING
Date Ackn.

OUTGOING
Date sent to post
Date receipt received

LOANS | Period of loan:

No. of Sheets Returned	Date	No. Outstanding

Reminder sent:

Acknowledged:

Extensions granted: *

Date H / 280 / 89

* LOAN GIFT EXCH TO FROM **

Name & Address: | Herb. Code

Region: | Bundles /
No. of Spms:
Family:
For attention of:
1.
2.

ACCESSIONS
ACCEPT LIST NAME(S) CHANGES

FIG. 14 — Suggested design for a tag. * = encircle as required; ** = add name or code of institution here.

required action has been taken, the tag is filed under the tag number. This basic plan can be elaborated further depending on the needs of the institution. One careful design of tag can cover all categories of consignments eliminating the need for further documentation, e.g. for accessions and duplicates, to be sent out (see fig. 14). Loans, however, will still need detailed listing.

Accessioning individual sheets

Many herbaria give each sheet a unique accession number and cross-reference it with a card catalogue including such data as: name of species, collector's name and number, country in which it was collected, etc. Such catalogues (when kept up to date) provide the following:

- An easily accessible summary of what may be found in the herbarium; this may save unnecessary handling and consequent deterioration of the sheets.

- A back-up record in case any sheets are lost while on loan or temporarily misplaced in the herbarium.

However, the labour involved in the compilation and maintenance (especially keeping the specific names updated) of such a catalogue is high, and since the herbarium is itself a catalogue this expense cannot always be justified.

Computerized databases (catalogues) are very effective as all data are rapidly accessible without laborious cross-referencing procedures (see **29. Introduction to computers**). It may well be worthwhile considering the adoption of such a system for new or small herbaria, but the labour involved in gathering and computerizing the data already accumulated in a large herbarium would probably be prohibitive.

Accession numbers may be added to the unmounted sheets as separate labels or stamped directly on the mounted sheet. In either case, care must be taken over their design and positioning so that any possibility of confusion with the collector's number is avoided (all too often the incorrect number has been cited in literature). A few herbaria (e.g. US) accession their specimens with bar codes as used in large supermarkets.

Shipping forms

These forms are used for keeping records of specimens in transit. They should be completed in duplicate with the details of the consignment. One copy is retained by the dispatching institution, while the other is enclosed in the parcel of specimens and then returned by the recipient herbarium signed as confirmation of safe receipt. Both forms should be filed; future proof of arrival may well be required (especially in the case of returned loans where reminders for overdue loans (recall letters) have been sent out erroneously).

Also send by air mail an invoice or advice slip ahead of the consignment, so that the recipient is alerted; this can be a further copy of the shipping form. This is useful in two ways:

- The researcher will know in advance details of what material to expect.
- If the specimens are lost in transit the institution expecting delivery will be able to inform the dispatching herbarium that nothing has arrived. The receipts of postage/shipment can then be used to trace the consignment.

ACCESSIONING, RECORDING AND DISPATCHING SPECIMENS

Incoming specimens

On arrival, all consignments of specimens should be taken to a reception office which will be apart from the rest of the herbarium (see **4. Pests and treatments**). Acknowledgement of the receipt of all parcels, boxes etc. should be promptly made to the dispatching institution or individual. Preprinted cards or proforma letters are useful for this purpose. When the contents have been checked, the shipping forms from the sending institution can be signed and returned. The parcels must then be treated for pest infestation, either by fumigation or deep freezing before they enter the main herbarium buildings.

Consignments of specimens may be divided into two broad categories: acquisitions and loans.

Acquisitions

Consignments of specimens will have been acquired in one of four ways:

- Collections by staff members on field expeditions.
- Duplicates received from other institutions in continuation of exchange (see **18. Duplicate distribution**).
- Gifts – either freely given (e.g. as complete private herbaria disposed of by the heirs) or donated in exchange for the identifications.
- Purchases – historically either collections (especially important private ones) were bought to build up the size and quality of a herbarium, or the services of professional collectors were sought. However, today very few institutions have sufficient funds for such purchases.

Whatever the means of acquisition all consignments must be accessioned and records kept of details that may be needed for future reference. At least the following data should be noted:

- Name and address of sender.

- Date of receipt.
- Status, e.g. gift, exchange, staff collection, purchase.
- Country in which the collection was made, and name of collector.
- Number of specimens (or sheets, as some specimens may be represented by more than one sheet).
- Type of plant, e.g. ferns, fungi, grasses or miscellaneous.
- Whether or not a determination list is required, and if so is there a time limit (e.g. priority naming).

The sorting of the consignments into families must be organized. In a large general herbarium a botanist with the appropriate knowledge of the geographical area concerned should be approached. Occasionally consignments of duplicates from more than one area may be received, in which case a geographical sort will first be necessary.

At this stage the quality of both the specimens and the data should be monitored and any necessary correspondence with the sender dealt with (e.g. it may be necessary to request data labels for some specimens).

If the consignment is to be split into batches for distribution to different botanists, careful records must be kept (especially if identification lists are required). Clear instructions must be passed to the botanists together with the specimens. These are best written on hanging labels and stuck to the bundles.

Incoming loans (see fig. 15, also **19. Loans to other institutions**)

Loans *returning* to the institution – these must be checked against the lists prepared when the specimens were originally processed for dispatch. It is important to confirm that the total number of specimens marked back is the same as that stated on the original form. Any discrepancies should be noted and the borrowing institution informed.

Loans *arriving* from other institutions – each sheet must be individually recorded. This can be done by writing on the sheet, in pencil, the tag (or accession lot) number beside a running number for each sheet. The running numbers are then listed against the details of the specimen (collector's name and number and name of species).

Outgoing specimens

Again, consignments of specimens can be divided into two categories: gifts and exchange and loans.

Gifts and exchange (see **18. Duplicate distribution**)

It is important to record the number of specimens sent, the recipient institution and date sent. Also include a brief description of the content, e.g. families, genera, geographical area of the specimens represented. Some institutions list all the duplicates they dispatch, or at least the type material. Decide if the information being recorded is really needed. If a strict exchange policy is maintained, it is necessary to compare how many specimens have been sent against those received for a particular institution. A request may be received for a label noted as missing from a specimen in a previously dispatched consignment: can this specimen be traced through the records? Remember to include such items as photographs and reprints with gifts and exchange, and make the necessary records.

Outgoing Loans (see fig. 15)

Loans being *dispatched* to other institutions (see **19. Loans to other institutions**). It is important to record loans carefully; the following data must always be noted:

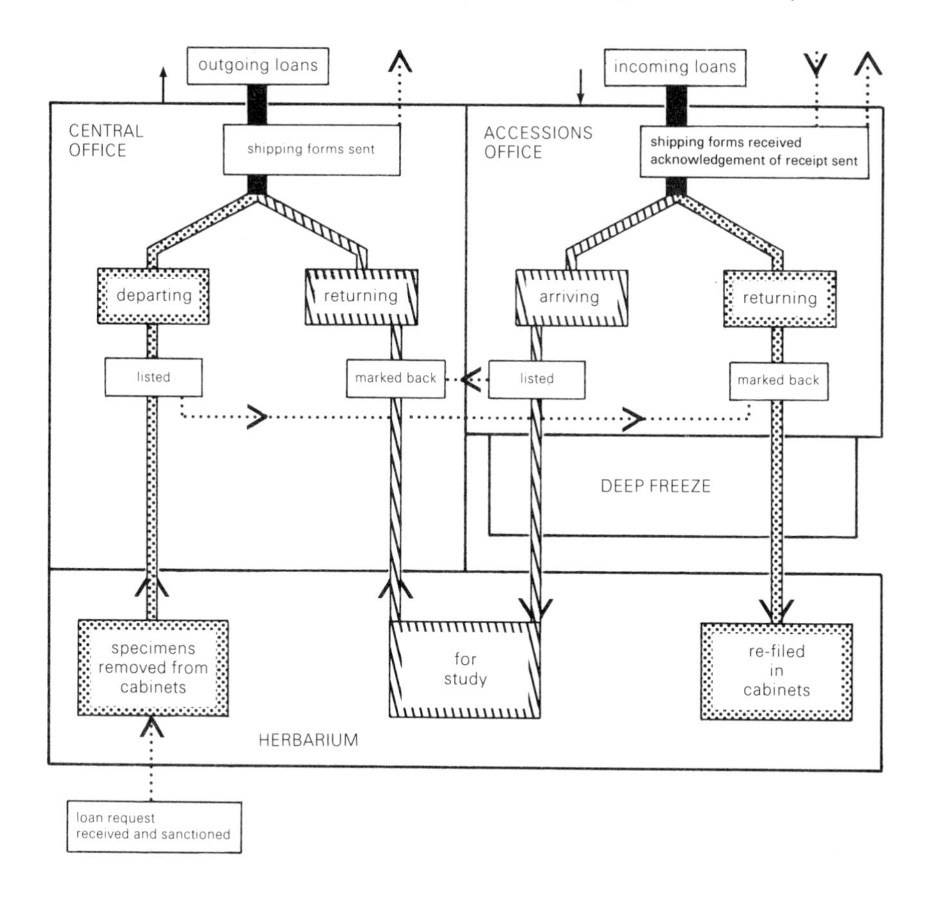

FIG. 15 — Flow-diagram of loan procedure. Solid black lines = incoming and outgoing loans; spotted lines = material belonging to home institution; striped lines = material belonging to other institutions; fine broken line = documentation.

- Institution and researcher to which the material has been sent.
- Date of dispatch.
- Date of arrival at borrowing institution.
- Period of loan.
- List of contents (so the material can be easily described and identified).

The records made when preparing the loan should provide the following:

- A list against which the loan can be checked when it is returned.

- Sufficient data to enable the recall letters to be sent when the loan period has expired.

Remember that a researcher is probably working on a number of loans from various herbaria at the same time. The official institutional stamp, consignment number and running number (see **19. Loans to other institutions**) should ensure that specimens are recognized and the consignment returned intact.

Enclose a list of the institution's regulations, so that the researcher is made aware of what he can and cannot do with the specimens.

Loans *returning* to other institutions:

- Carefully check that the material is in order. Erase all pencil annotations that may have been added when the loan was marked in.

- Check each specimen against the record made when the consignment arrived.

- Inform the lending institution if the complete loan is being returned or only part of it. Indicate what is being retained for further study.

CENTRALIZED CORRESPONDENCE

- Recall letters or reminders for overdue loans should be sent out on a regular basis, usually annually. All loan records should be checked and any overdue loans noted and letters (or proformas) dispatched. If requested, extensions may be offered providing this has been approved by a senior botanist. A note of recall or a reminder should be added to the loan record to provide a full and updated account. If another worker urgently requires to see material that has been sent on loan a recall letter must be sent, subject to the advice of a senior botanist.

- If duplicate specimens have arrived without data labels, letters requesting the data must be drafted and dispatched.

- Determinations provided for gifts must be recorded and sent to the donors. Incoming determinations must be recorded and correlated with the relevant specimens.

- If an accessions catalogue exists, great care must be taken to keep it up-to-date. Enter all new determinations and name changes, not forgetting to note any on returned loans. If the catalogue is not constantly updated, institutions which generate loan lists by computer for outgoing loans by entering accession numbers or bar codes will be listing erroneous information, and time will be lost searching for specimens in the wrong place.

PACKING AND ASSOCIATED PAPERWORK

In tropical countries where insect infestation is common, it is recommended that specimens are subjected to decontamination (see **4. Pests and treatments**) before they are packed for dispatch. Transit can take several months during which time insects can cause a lot of damage. Visual inspection is not sufficient as eggs can hatch in transit.

Care should be taken when preparing a consignment of specimens for dispatch. Mounted specimens should be protected by placing inside folds of paper (double flimsies or newspaper) and held securely together in blocks. This can be done by bundling between strong but light boards or wrapping in paper and sealing. The bundles should then be packed in a strong box along with information for addressee. A wrapping of waterproof paper should be added to protect against getting wet in transit. Labels and customs forms must be securely attached.

Paperwork to satisfy national Post Offices and Custom Offices vary from nation to nation. However, ways of advising of delivery and noting contents is uniform. If possible receipts of postage or shipment should be obtained. Type material should be registered or insured.

Over holiday periods (e.g. Christmas and New Year) there are likely to be delays in the postal services of many countries. For the safety of specimens it is advisable to try to avoid sending parcels when they are likely to be held up in such delays.

REFERENCES

Fosberg & Sachet: 97–100 (1965)
Millspaugh (1925)
Womersley: 70, 71 & 94 (1981)

8. PROCESSING UNMOUNTED SPECIMENS

It is preferable that all specimens are identified before being mounted. Unmounted material is easier to place under the microscope, to examine from both sides and for the removal of flowers or fruit for dissection.

If a *mixed gathering* has been noted when the specimen was named, the taxa should be separated and a copy of the data put with each taxon. The specimens must be renumbered with the suffix A or A and B (see **30: Collecting and preserving specimens**, *Numbering the collections*).

Listing determinations

The determinations (scientific names) may be needed by other herbaria or individuals. Names should be listed with the collector's name and numbers in sequence. If the handwriting on the determination label is unclear the spelling can be checked in:

- Brummitt, 'Vascular Plant Families and Genera' (1992), Mabberley, 'The Plant Book' or Willis, 'A Dictionary of Flowering Plants and Ferns' for the generic name.
- 'Index Kewensis' for the generic and specific names and author.
- Brummitt & Powell, 'Authors of Plant Names' (1992) for the correct form of the authority.

It is important that the list is legible and preferably typed (fig. 16,A).

The listing procedure is computerized at some herbaria, but this may not always be possible. (See fig. 16,B).

A copy of the list should be filed for reference. Lists can be an invaluable aid for working out itineraries or forming the basis for check-lists.

When some herbaria (e.g. MO) distribute duplicates for naming they include specially printed slips on which the determination (or redetermination) should be noted. These slips are eventually collected together and returned to the sender so that the names can be noted and curatorial amendments made (see fig. 16,C).

Preparation for mounting

At a large herbarium usually some staff are employed solely as mounters, and it is necessary for the technical staff to check and prepare the specimens before they are sent for mounting. In a smaller herbarium it is possible that both preparation and mounting will be done by the same person, in one or two operations.

A

TAG NO. H1310/77
H341/78

ROYAL BOTANIC GARDENS, KEW

A PARTIAL LIST OF DETERMINATIONS AND NAME CHANGES

Collector: Mrs Jean Pawek

Sender: Mrs Jean Pawek, # 166, 191E E.Camino Real, Mountain View, California 94040, USA.
Checked by: R.K.B.

Country of origin: Malawi Area 10E
Date: 21 November 1979

1783	Streptocarpus solenanthus Mansf.
1830	Cynoglossum sp.
1985	Streptocarpus buchananii C.B.Cl.
1990	Streptocarpus solenanthus Mansf.

B

ROYAL BOTANIC GARDENS, KEW

List of Determinations

ADDRESS: The Administrative Curator,
Herbarium,
Missouri Botanical Garden (MO),
P.O. Box 299,
St. Louis,
Missouri 63166,
U.S.A.

DESPATCHED BY: JBJ Blewett

DATE: 11/04/89

H/315/88 Area: 10E MALAWI

Chapman & Chapman

8250	Lycopodium dacrydioides Bak.
8265	Asplenium megalura Hieron.
8266	Asplenium aethiopicum (Burm.f.) Becherer s.l.

C

MADAGASCAR

ANACARDIACEAE — L. J. Dorr with
Initial det — L. C. Barnett & A. Rakotozafy 4582
Rhus

New or updated det. Protorhus sericea Engler

Determined by: D.L. Forman — Date: 28.iii.89

Computer — Fieldbook — Mounted

If a new determination is made please return this slip to : Missouri Botanical Garden, P. O. Box 299 St. Louis, MO 63166-0299 U. S. A. Thank you.

PLEASE DO NOT MOUNT THIS SLIP

FIG. 16 — Examples of determination lists. **A** conventional typed list; **B** computer-generated list; **C** slip for individual specimen determinaton.

Laying-out (or the arrangement of the specimen on the mounting sheet) is dealt with under **9. Mounting herbarium specimens**.

Single specimens (i.e. sufficient material of a gathering for one specimen).

Check that the material is sufficient for one sheet (or 2 or more if needed (see below) to show the full range of material), that the label(s) is(are) present and any ancillary collections are cross-referenced (see below).

If the specimen is of poor quality or sterile, consider whether it should be kept or discarded (refer to senior botanist). There can be good reasons for keeping sterile specimens; before any are discarded consider the following:

- in many species flowers and leaves do not appear together in the same season. This often happens in woody plants, but also in herbaceous taxa such as *Equisetum*, many Aroids, some orchids (e.g. *Nervilia*) and many other petaloid Monocotyledons.

- coppice or sucker shoots frequently have foliage atypical of the species. The collection may have been made to demonstrate such variation.

- some sterile collections may have been made as they are known to be the only record of the taxon for the area concerned. This should be indicated on the data label, but it may sometimes be necessary to check.

Forceps can be used to handle the more fragile material and to transfer any loose items to the correct sized paper capsule; alternatively slide a small piece of stiff paper under the delicate item to be lifted. The capsule should have the collector's number marked in pencil. Any very delicate parts should be placed in a separate capsule from other items. If necessary, fasten the capsule with a no-tear or continuous loop paper-clip (see **5. Materials**).

Specimens already mounted when received may not be on the correct paper size and quality, and will need to be standardized to the herbarium's specifications (see, **10. Conservation of sheets**).

- Strapped specimens can easily be removed from the paper by disengaging the gummed paper straps with forceps, removing or cutting around the label. If 'Sellotape' has been used remove as much as possible without damaging the sheet.

- Glued specimens should be detached from the sheet if this can be accomplished without harming the specimen. Otherwise trim the sheet around the specimens and labels, then remount.

If needed, extra data labels for second sheets can be photocopied or if hand-written include only the essential data. 'Sheet 1', 'Sheet 2' etc. must be written (or rubber-stamped) on the multiple labels, preferably in red to attract attention (see fig. 17,A). If the labels accompanying the specimens are of poor quality, new labels should be made and the originals retained in paper capsules. Extra large labels should also be folded and placed in paper capsules and the essential data copied on a standard-sized label. Any type material should be clearly marked as such, preferably on a separate label (use special TYPE labels if available, see fig. 12). (Refer to senior botanist if in doubt as to the kind of type label required; see **23. Rearrangement of herbarium collections to new publications**).

If there is accompanying spirit, carpological, wood or photographic material, cross-references must be made. Use rubber stamps with red ink or write by hand, e.g. 'see Spirit Collection', 'see Herbarium Material' etc. and add a copy of the data (or at least essential data) to the ancillary material (see fig. 17,B). If a large fruit is still attached to the specimen it may have to be detached for transfer to the carpological collection. If possible, first photocopy the specimen to show the position and manner of attachment, otherwise record this data.

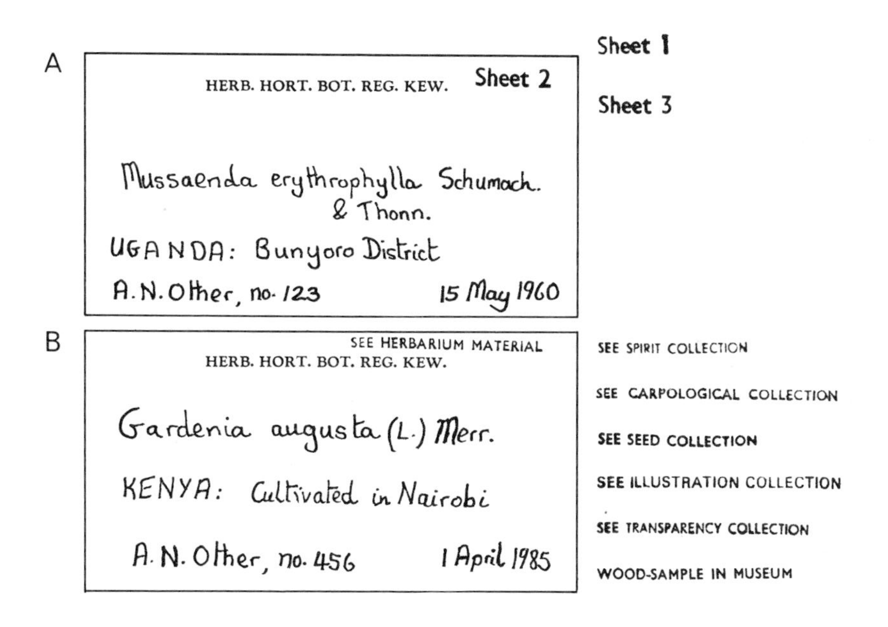

FIG. 17 — Examples of labels with essential data and appropriate rubber stamps (red ink preferable). **A** rubber stamps to indicate multiple sheets; **B** rubber stamps to indicate material in ancillary collections.

Specimens with duplicates to be extracted

These will mostly be specimens collected for one's own institution, but occasionally duplicates from other herbaria may be found to consist of more material than it is necessary to retain. Often the duplicates will have already been sorted into sets by the collector. When sorting duplicates, the following points should always be considered and checked:

- If the flowers are unisexual, both ♂ and ♀ flowers should be represented in each duplicate.

- If the flowers are of two forms, long-styled and short-styled (heterostyly), both forms should be represented in each duplicate.

- Some taxa are very variable and care must be taken to represent the maximum range or variation on each duplicate – e.g. smallest and largest leaves, shape ranges from round to elliptic, indumentum ranges from glabrous to pubescent or leaf arrangement changes from opposite to alternate. Note especially annual plants where entire plants can vary in the amount of branching and overall size.

- Care should be taken not to mistake heterophylly (i.e. two types of leaf found on the same plant) for mixed gatherings. Juvenile foliage, especially, can look very dissimilar to the adult leaves. Give each duplicate examples of both leaf types.

- As far as possible make sure each duplicate has a complete range of stages – flower buds, flowers, young fruit, mature fruit.

- If a parasitic plant has been collected together with a sample of its host, take care that the two are either kept together or cross-referenced.

It is as well to remember that sterile portions will in general not be welcome as duplicates in other herbaria. If sterile material is sent out it should only be done so for a good reason, see above, *single specimens*, and this reason indicated so that the herbarium receiving it does not discard it as substandard.

Bearing the above points in mind, duplicates can be extracted and a set of data inserted with the material together with an 'EX HERB' slip (e.g. 'EX HERB. HORT. BOT. .....' or 'FROM THE HERBARIUM OF ......') (see fig. 9, F&H) or cancellation stamp. If the botanist has indicated duplicates of holotype material, include '*isotype*' stickers. (See **23. Rearrangement of herbarium collections according to new publications**.)

Each duplicate is placed in a double flimsy or newspaper; if the material is bulky, place a second flimsy the opposite way round. The duplicates can then be distributed to other herbaria. (See **18. Duplicate distribution**). Specimens to be retained can be stacked in their original newspapers or double flimsies and sent for mounting.

REFERENCES

Brummitt (1992)
— & Powell (1992)
Fosberg & Sachet: 64, 66–69 (1965)
Mabberley (1987)
Willis (1973)
Womersley (1981)

9. MOUNTING HERBARIUM SPECIMENS

Check-list of materials (see **5. Materials** for specifications)

Cartridge paper or white board (for bulky specimens)
Approved adhesive for applying labels and specimens
Paper capsules
Translucent or transparent packets
Paper-clips
Linen or cotton thread
Gummed paper, archival quality if possible (for backing stitches)
Gummed linen tape or archival self-adhesive tape (for strapping)
Fine translucent paper (windows)
Waxed paper
Drying paper
Transparent polyester sleeves, slightly larger than the sheets

While unmounted specimens are best for the purposes of examination, dried material tends to be very brittle and susceptible to damage. However, properly prepared and well cared for mounted herbarium specimens will last indefinitely. It is therefore essential to select mounting paper, adhesives etc. of archival quality. The two main aims of the mounting process are:

1. To display the specimen and data to allow maximum observation (laying-out).

2. To preserve the specimen by securely attaching it to strong mounting paper or card, but at the same time allowing for the removal of small portions for more detailed study.

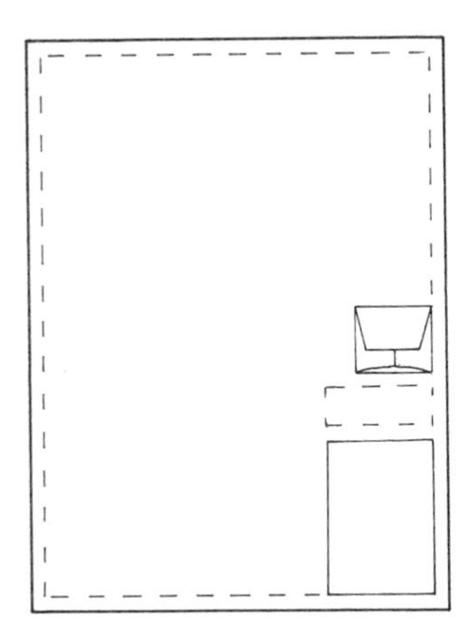

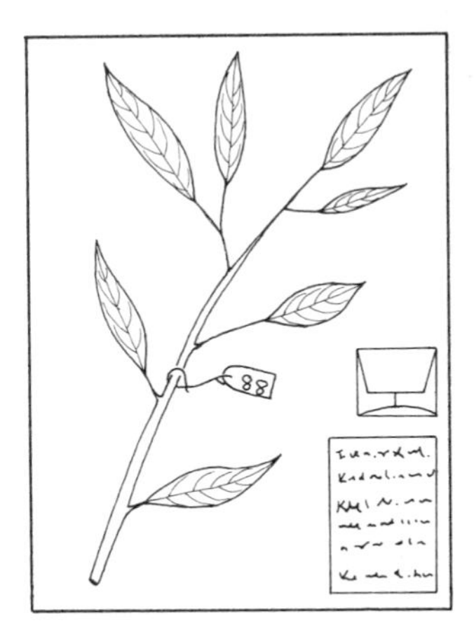

FIG. 18 — Mounting sheet with labels, capsules and specimen positioned free from margin.

LAYING-OUT (OR ARRANGEMENT)

Choose the best way of presenting the specimen in order to display all its characters as far as possible, and not just to show the most pleasing or artistic arrangement. Consider the best position for the labels and capsule before attaching them to the sheet. There are good and bad areas to place items. The extreme edges of the sheet should be avoided; when mounting try and maintain an imaginary border around the sheet (see fig. 18). Also bear in mind that when in the genus cover, more pressure is likely to be exerted on the left hand edge of the sheet.

Labels

It is often best to consider the position of the label first, but to attach it last. The best position for the main label is generally thought to be the bottom right; this makes the label easier to read when kept in genus covers which open on the right.

Additional labels should be placed (if possible) above the main label or at least close to it. Ideally a space should be left above the label to allow for the future attachment of determination slips (see fig. 18). The label should be stuck down completely. If this cannot be done because the specimen is too large, attach the label by one edge only (the outer vertical is preferred). Make sure that the label is not lying across a bulky or projecting part of the specimen which may damage it. Always check the reverse side of labels, as there may be maps or further information, in which case attach by one edge only.

If it is not possible to mount the label in the bottom right hand corner, it can be mounted elsewhere on the sheet. Very large labels may be folded and stored inside paper capsules.

Paper capsules

Choose the correct size of capsule to contain the loose portions concerned.

If possible place the capsule on the right-hand side of the sheet corresponding to the open edge of the genus cover. Avoid placing it on the left as this causes excessive thickness in the fold of the genus cover and this can lead to damage.

If a particular taxonomic group of specimens is likely to acquire bulky capsules, vary the capsule position to prevent a wedge-shaped stack of sheets.

Translucent packets containing such items as small seeds should be placed unsealed inside the paper capsules. If flowers are plentiful, always consider placing some in the capsule to ensure that they can be easily studied.

The number tag

This is a tag bearing the collection number of the specimen; it is usually attached to the specimen by thread (or sometimes a piece of paper attached to the specimen may have been used). It must be preserved with the specimen by glueing to the sheet either:

- beside the main label,
- or still attached to the specimen if it does not obscure any part of the specimen.

Specimens

Removal of duplicates, bulky items and multiple sheets (see **8. Processing unmounted specimens**).

If the specimens have been very poorly pressed it is possible to improve them by soaking overnight in water containing a wetting agent, rinsing with clean water, rearranging the specimens and repressing in a plant press until dry. The drying time will be considerably less than for fresh material.

Arrange the specimens, bearing in mind the following:

- Choose the best side to display as many features as possible.

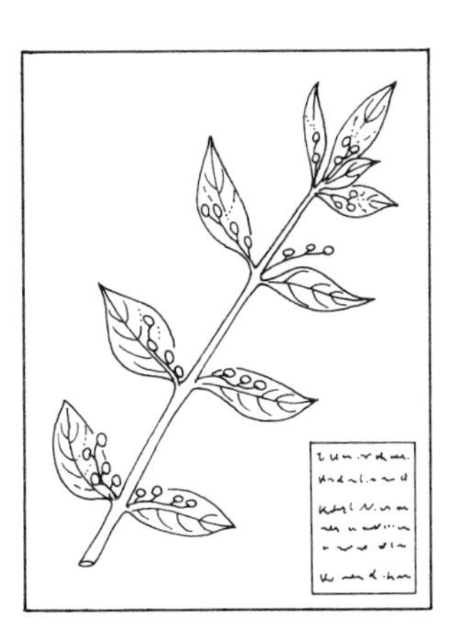

– Expose hidden flowers or fruits by removing leaves and placing into paper capsule; together with any other loose items.

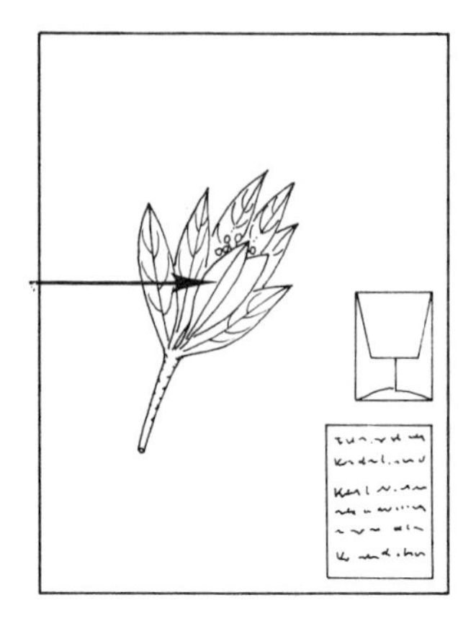

– Display both sides of leaves – if necessary, detach and turn one leaf or place in capsule.

– If only one large leaf, cut off part and turn or place in capsule.

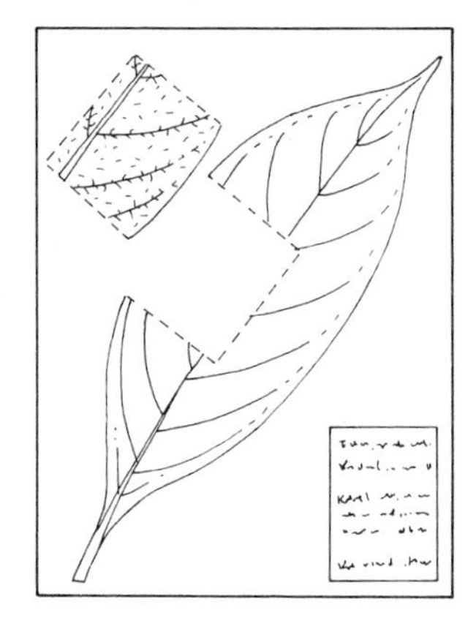

— Display both aspects of the flower where possible.

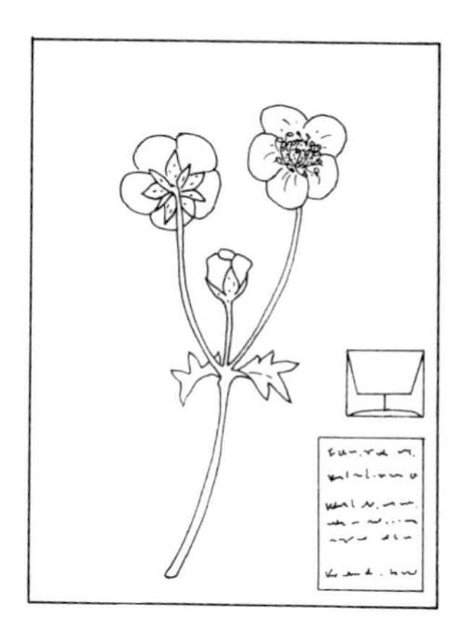

— If possible separate clumps of plants without damage, carefully removing any soil from roots.

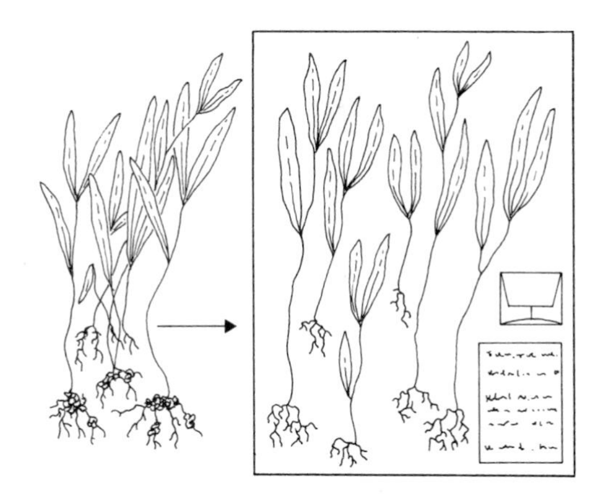

– When mounting more than one plant on a sheet:

1. keep them all aligned the right way up.
2. place the largest or heaviest specimens at the bottom, to prevent the sheet from bending when handled.

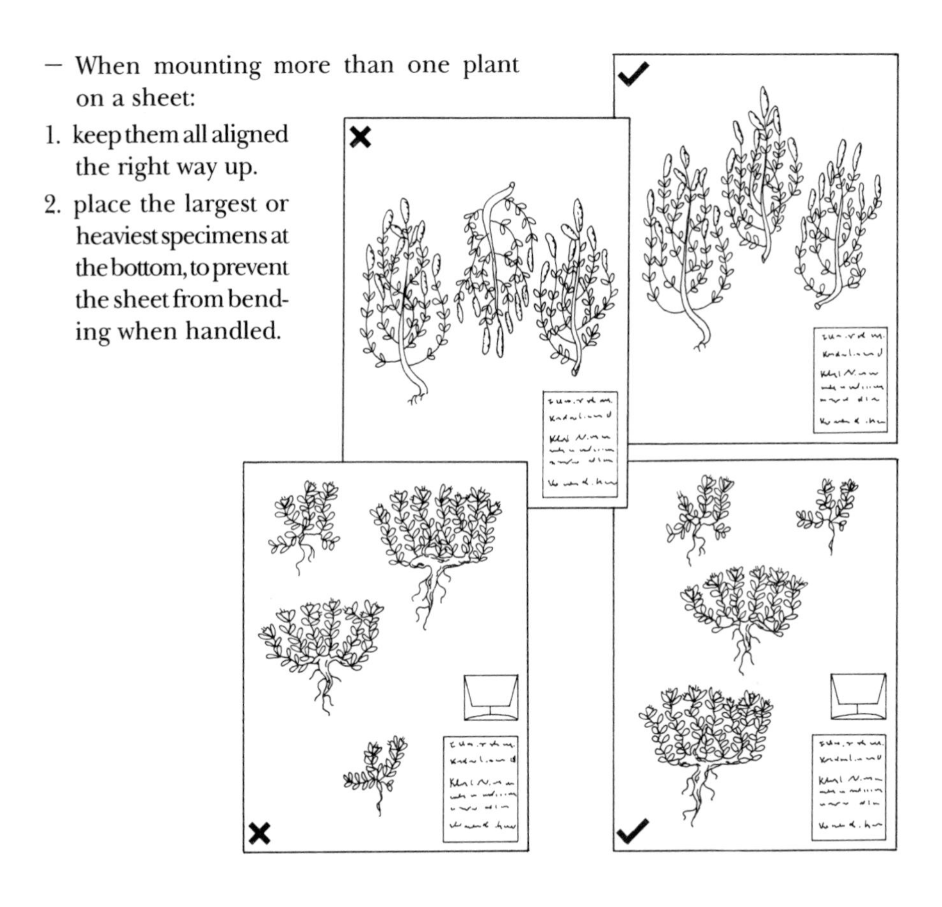

– Tiny plants: if numerous, spread out a few, but place the majority in a capsule; if only a few keep all in the capsule.

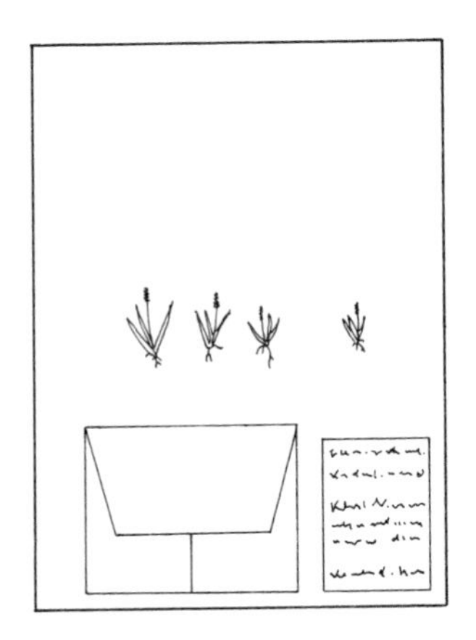

— Trimming over-large specimens:

1. trim just stem if possible.
2. only trim leaves if plenty of good leaves are left.
3. remove a whole leaf and place in paper capsule.

— Bulky specimens: place on white board. Trim lumpy specimens with forwardly projecting spines or branches which could cause damage to adjacent sheets in the herbarium.

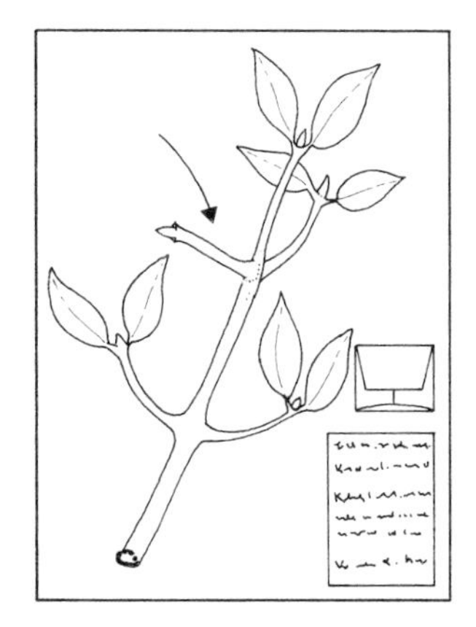

— Avoid placing bulky or delicate items near the left side of the sheet, where they can get crushed in the fold of the genus cover.

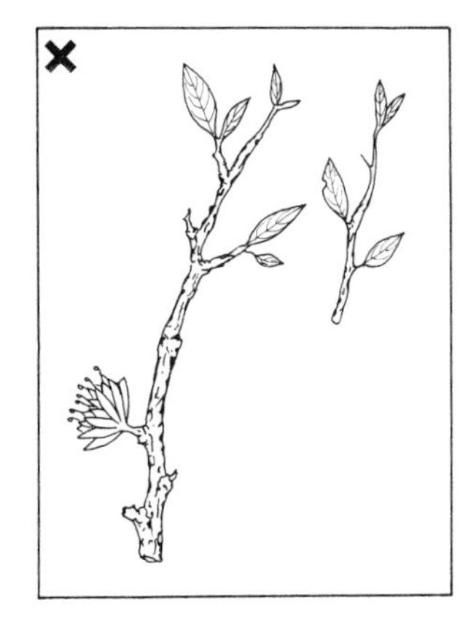

– Many larger specimens are best arranged diagonally. This provides both more length and width than positioning longitudinally; it can also prevent parts of specimens lying behind the label.

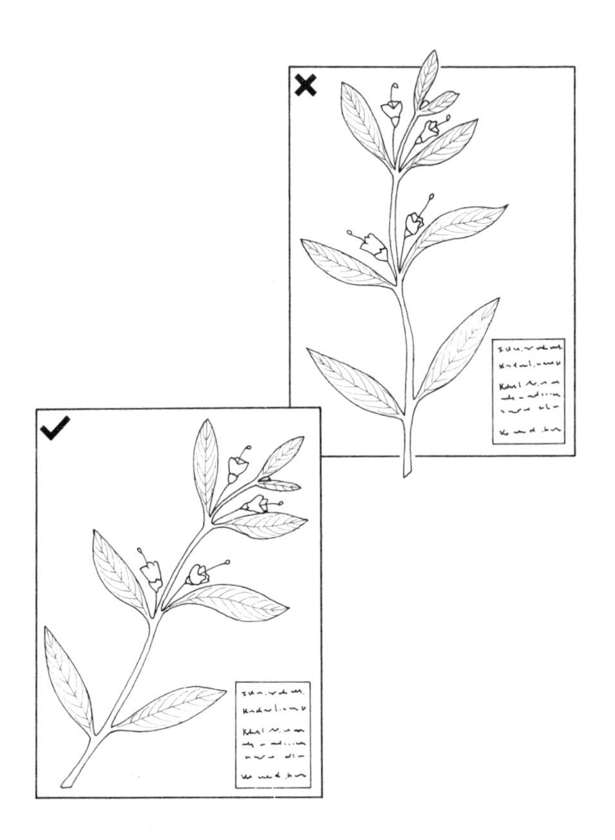

— Over-long specimens can be folded to fit the sheet so that the apex points upwards or the base downwards.

— If a specimen has only one large flower this should not be stuck down. However, it must be protected by placing a 'window' of translucent paper over it. This window is then secured by its outer edge only, allowing it to be folded back for close examination of the flower and its point of attachment to the specimen, information which is lost if the flower is detached.

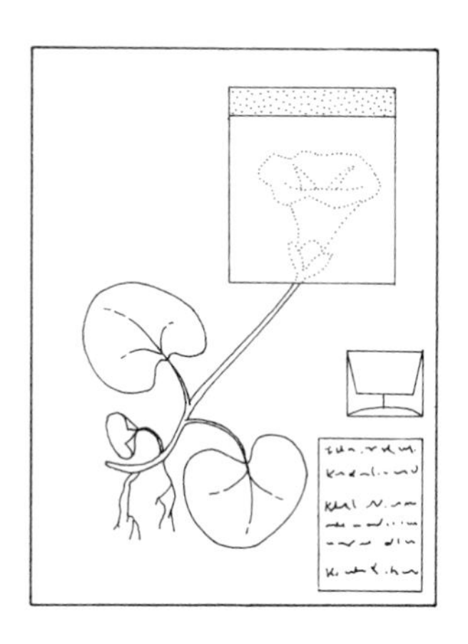

— Specimens with inflorescences which tend to fragment (e.g. Gramineae) can be protected by placing the whole sheet in an unsealed transparent sleeve. Note: inflorescences of Gramineae specimens should not be stuck to the mounting sheet.

MOUNTING (methods of attaching the specimen)

There are two main ways of mounting specimens: *strapping* (the 'straps' may be either thread, linen tape, archival self-adhesive tape (various widths available) or plastic glue) and *overall glueing*. There are arguments for and against both methods.

STRAPPING	OVERALL GLUEING
Easy to reverse (should remounting be necessary).	Can only be removed if reversible (water soluble) glue was used.
Will allow easy removal of portions for detailed study.	As full a range of organs as possible must be placed in a paper capsule before mounting.
Decontamination by deep freezing (see Egenberg & Moe 1991) or heating should not cause damage by differential contraction or expansion between the differentially hygroscopic specimens, paper and glue.	Damage by differential contraction and expansion is possible, especially to delicate flowers. These, however, are normally left free and 'windowed' (see below). Handmade and mould-made paper will cause less damage.
Specimens are susceptible to damage as this form of attachment still allows a certain degree of movement when the sheets are handled, or sent through the post.	Specimens are firmly attached to the sheet, thus reducing damage.
Fragments can easily be removed illicitly.	Helps to guard against the illicit removal of fragments.
The labour involved in any of the strapping methods can be greater than in glueing.	If glueing is used, bulky specimens may require some reinforcement by strapping, usually with thread.

The decision as to whether to mount by strapping or by overall glueing can be difficult. Strapping is perhaps appropriate for small herbaria with restricted access to visitors, while overall glueing can give better long-term protection to specimens in large, busy herbaria with free access to visitors and frequent requests for loans.

Strapping methods

Sewing

Stitches made with linen or cotton thread must be regularly knotted off on the reverse side of the sheet. Double the stitches on stems to give added strength. The knots should be covered on the reverse side of the sheet with strips of gummed paper to prevent the thread from catching on underlying specimens when in the herbarium. A few herbaria knot off on the face of the sheet.

Strapping

Small straps of gummed linen tape are placed across the specimen at intervals and stuck to the sheet at either end. Thin straps should be used on finer parts of the specimen and thicker straps where extra strength is needed, e.g. thick stems. Archival self-adhesive tape has only recently become available and is at present on trial in some institutions.

Avoid strapping over important details, such as flowers or small inflorescences (see fig. 19). The aim is to hold the structure firmly but keep it free for examination.

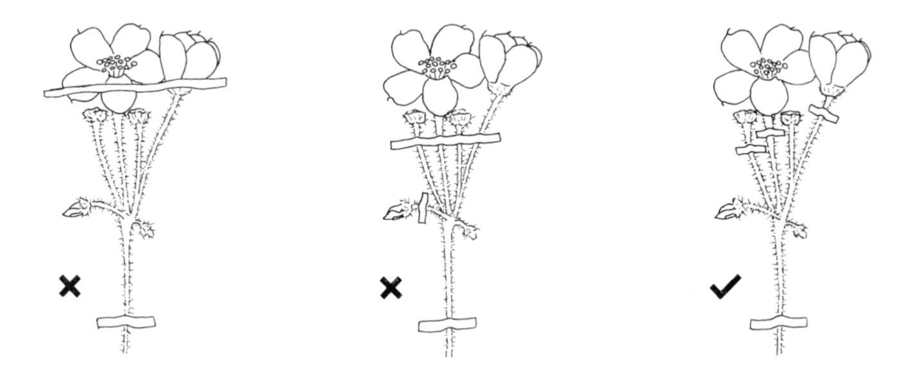

FIG. 19 — Examples of incorrect and correct positions of straps. The straps should be short and cover only one pedicel; they should be near the flower but not passing over the petals.

An exception to this is where there are very long, delicate inflorescences which obviously need extra support to prevent damage, e.g. Gramineae. Here the straps should be placed near the base and at convenient points along the inflorescence (see fig. 20).

When strapping is applied across a leaf-blade, only attach the ends of the strap where contact is made with the sheet. This allows the leaf to move freely when the sheet is handled.

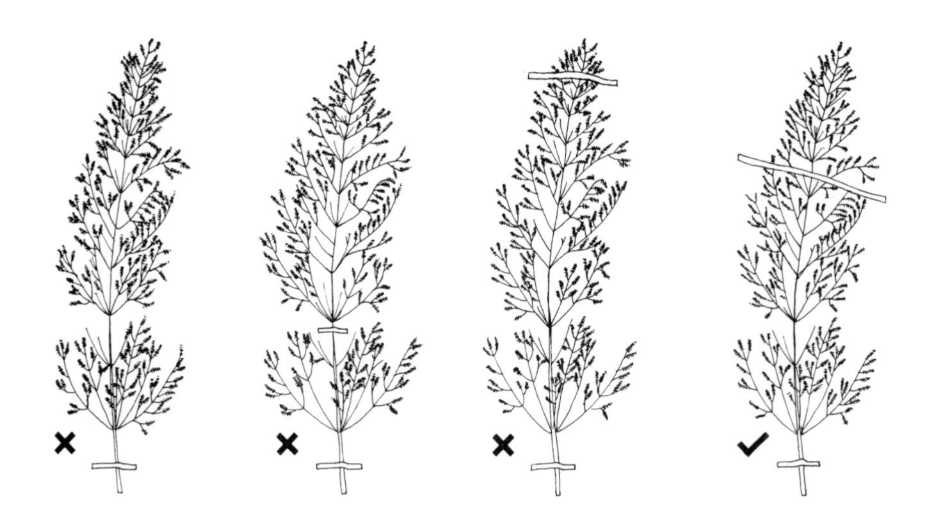

FIG. 20 — Correct and incorrect positions of straps for grass inflorescences.

The Archer Method

This method of mounting was devised by Archer (1950). Using a nozzled applicator a thick, liquid plastic adhesive is placed across parts of the specimen at intervals like a series of straps and allowed to dry. The plastic straps which result are very strong (see p. 39).

The adhesive can also be applied as spots at the edges of leaves to attach them firmly to the sheet after the specimens have first been attached to the sheet by overall glueing with a light glue (fish glue was formerly used). This method is sometimes referred to as 'blobbing'.

Overall glueing

Adhesive is applied to the reverse side of the specimen so that virtually all the specimen is firmly attached to the sheet.

Water-based woodworking adhesive (e.g. 'Evostick resin W'), library pastes and latex adhesives are often used. However, archival quality PVA glues which are more easily reversible are now available and are preferable, as is methyl cellulose (see **5: Materials**, *Adhesives for mounting*).

In some cases delicate leaves and petals will absorb the water in the adhesive and curl up. It is best if these are avoided and only the stems glued, especially if decontamination by deep freezing is to be anticipated. The specimens can be protected by 'windows' or by placing inside a transparent sleeve.

It is best to work quickly and apply the glue in one action, but avoid using too much. Any of the following methods can be used:

- By brush. This can be quick (especially important with quickly drying latex), but watch out for easily detachable fragments of specimen attaching to the brush and messing up the glue. These fragments may later come off on other specimens (or labels). It is common practice to use separate glue pots and brushes for the labels and specimens.

- Spread the adhesive on a plate (ceramic tile or glass plate). Place the specimen on the adhesive then pick up and attach to sheet. This is a useful technique when dealing with small or delicate specimens with fine stems.

- Applying directly to reverse side of specimen by squirting from a nozzled applicator. This can be quick, and the lack of contact with the specimen avoids picking up fragments. However, the nozzle will require frequent cleaning.

Material prepared by overall glueing must be left under pressure while the glue dries. This is done by covering each mounted sheet first with waxed paper and then with a sheet of drying paper. The procedure is repeated until a pile of protected specimens is built up; it is then weighted down with sandbags and left overnight.

The specimens can then be 'finished'. Rigid or bulky stems which might tend to lift free should be stitched firmly to the sheet (see above, *Sewing*). Adhesive should be applied to parts which are still free by sliding under leaves etc. with a flat knife blade. If transparent windows are necessary they should be attached at this stage. Re-glued specimens should again be left under pressure while the glue dries.

Aquatic plants

The specimen is likely to take up the water from a water-based glue and become distorted, unmanagable and difficult to mount. For this reason latex glue has been used for mounting delicate aquatic angiosperms. Discoloration may also occur.

The following method should be adopted:

1. Carefully arrange the specimen on a mounting sheet.
2. Cover with a sheet of waxed paper.
3. Without disturbing the specimen turn the sheet + specimen + waxed paper over as a block so that the waxed paper is now at the bottom.
4. Remove mounting sheet, and apply glue to the area which will cover the specimen.

5. Replace mounting sheet glued side down.
6. Invert the stack again.
7. Rub down the specimen through the waxed paper in order to ensure that the specimen is in contact with glue. Remove the waxed paper carefully. Some parts may need extra glueing.

The finished sheets can be stacked with waxed paper, drying sheets and sandbags as described above.

Attaching labels and paper capsules

For the sake of convenience, labels and paper capsules can be attached with the same glue as the specimens. The glue can either be blobbed close to the edges or applied evenly by spreading on a plate (see above). Some labels, especially if made from paper with a grain, tend to expand as they absorb moisture from the glue and can, if not handled carefully, wrinkle. To a certain extent wrinkling and lifting of the corner of the mounting sheet will lessen as the specimens dry under pressure.

REFERENCES

Archer (1950)
Croat (1978)
Egenberg & Moe (1991)
Fosberg & Sachet: 69–76 (1965)
Savile (1973)
Womersley: 72–80 (1981)

10. CONSERVATION OF SHEETS

The paper on which old herbarium specimens are mounted can become fragile with age (especially if lightweight or thin), torn at the edges and often very dusty. Also the glue tends to deteriorate and the specimens and labels can become detached. In the past, it was common practice in some herbaria for more than one specimen to be mounted on a sheet (i.e. **mixed sheet**). Sometimes undetected **mixed gatherings** may have been mounted together. Although such specimens cause curatorial problems, considerable thought is necessary before correction is attempted. It must be remembered that a herbarium specimen is a source of information, and that apart from the plant itself, information can also be contained in the labels (including determination slips and 'seen for ...', stickers), drawings, notes, handwriting, rubber stamps, paper capsules and their content, and even the kind of paper of the sheet itself. Such items supply information (or clues) as to the history of the study of the plant: who studied it, when, and with what conclusions? In view of this, cleanliness and tidiness, although desirable, must be secondary considerations.

CLEANING SPECIMENS

If the Herbarium building is not air-conditioned and the cabinet doors not air-tight the specimens will become dusty. Some dust, e.g. from unsurfaced roads, may be relatively inert, but dust of industrial origin can contain harmful acids. It is desirable that dust be removed, if time permits. Some people are sensitive (allergic) to household dust; they should take the precaution of wearing a light disposable mask when disturbing the dust on herbarium specimens.

The sheets should first be dusted with a soft-haired brush (a thick watercolour brush or camel hair brush) to remove loose particles. If the leaves and stems of the specimen are robust they can be similarly treated, but flowers and other delicate structures must be avoided for fear of damage. Any remaining dust on the sheet can be removed with the help of a slightly adhesive substance such as '*Blu-Tack*', a putty rubber (kneaded rubber) or a special archival product (e.g. '*Absorene*'); the use of breadcrumbs is not recommended, especially since any residue may attract insect pests or moulds. The adhesive substance should be lightly dabbed on the affected area (*not* rubbed on), avoiding the plant specimen; it can be rolled to a point so that the specimen can be approached more closely. It is possible to clean labels in this way, even pencil writing will not be removed, *providing* they are not rubbed. Never use an ordinary office eraser for this purpose.

Stains on herbarium sheets such as those caused by the application of liquid insecticides or dampened non-waterproof ink are best ignored; they do not detract from the scientific use of the sheet. However, small rusty-brown spots known as '*foxing*' (see **5. Materials**, *Paper glossary*) can indicate that the sheet is of a poor quality paper; 'heavily foxed' sheets should be replaced (see below, *Re-backing & re-mounting*). See CCI (1990).

REPAIRING SPECIMENS

Loose portions are always best dealt with as soon as noted. If the portion or item has completely broken free, place in a paper capsule, which is then glued to the sheet. If the portion is loose or free at one end, either introduce new glue under it using either a palette knife or strip of stiff paper, thus reducing damage by bending, or re-strap or re-stitch as appropriate. Loose labels should be re-glued in their original positions.

Fragile or faded labels can be replaced with a newly copied version, but the original *must* be preserved in a paper capsule mounted on the sheet. The outside should be annotated 'original data label'.

Minor tears can usually be mended with strips of acid-free bond paper cut to size and glued with the approved herbarium glue; gummed tape (c. 2 cm. wide) can be used providing it is of archival quality. Sometimes numerous minor tears can indicate poor or deteriorated paper; if this is thought to be the case, see below, *dealing with whole sheets*. There are three main types of tear:

– *damaged margins* – numerous splits usually not extending more than 1 cm. into the sheet. Place paper strip or tape along the damaged edge so that it projects slightly beyond the sheet; ease down the edge of the sheet and smooth with a bone-folder or other flat smooth implement; trim tape with a scalpel and steel ruler when glue is dry. If the sheet is larger than standard, the margin may be trimmed *providing* none of the labels, writing on the sheet or specimen is lost.

– *short tears* extending up to c. 7 cm. into the sheet. Glue paper strip to underside of sheet along the length of the tear; smooth down sheet from above; trim at edge when dry.

– *puncture holes* caused by upwardly projecting parts of the specimen below in the species cover. On the back of the sheet stick sufficient paper to cover the hole and surrounding area; from the front flatten and smooth down the edges of the hole with a blunt smooth instrument (a rounded pen top would do), the torn edges should join together again leaving a smooth surface.

Dealing with whole sheets

If the original mounting sheet is brittle and too thin (or lightweight) to provide adequate support for the specimen, or is very damaged, it may either be placed in a protective cover or remounted:

– *Protective covers* – the type of transparent sleeve (see **5. Materials**, *Transparent sleeves*) used to protect grass specimens is ideal for all but the most delicate specimens where removal from the sleeves for study could cause further damage. If these are not available, type covers could be used, but they must be annotated – 'NOT TYPE MATERIAL'

- *Re-backing sheets* – these may be mounted as a whole on a new sheet. First check to see if anything is written on the back of the sheet; if it is, photocopy (see below) and place the copy in a paper capsule to be mounted with the sheet. If the margins are excessively damaged, they may be trimmed, providing nothing is written close to the edge and the labels are not cut. The labels should be left in their original position; they can be re-glued if necessary, but if fragile, see above, *fragile or faded labels.*

- *Re-mounting* – very occasionally it may be necessary to replace the original mounting sheet. For example the sheet may become brittle and discoloured due to acidity in the paper and this could lead to the deterioration of adjacent sheets by acid migration. First check for writing both on the front and back. Photocopy if possible, but faint wording in pencil or coloured rubber stamps may not show in the copy, in which case writing and stamps must be cut out and saved. If the labels are fragile see above, *fragile or faded labels.*

  First draw an outline in pencil, then cut away as much as possible of the old sheet from around the specimen. Although scissors can be used if the plants are well spaced, for the best results place the sheet on a soft board and using an artist's cutting knife or scalpel cut as close to the specimen as possible without risk of damage. This method will leave smoother edges and avoids any bending of the specimen. The specimen with its paper backing can then be glued to a new mounting sheet, using the approved mounting adhesive, covered with waxed paper and placed under a sandbag until dry. The greatly reduced area of the old mounting sheet now backed will reduce acid migration.

  If the specimen was mounted with water-soluble glue it may be possible to lift it off the sheet after soaking in water, warm if necessary. This can be done by placing it on a wet pad of absorbent paper or cloth for a few hours. If there is the possibility of the writing on the labels not being waterproof, these should be cut out beforehand; typewritten labels can be soaked off.

  If the specimen was strapped or stitched it is relatively simple to remove it from the sheet. Use forceps to disengage gummed paper straps or cut thread with either small sharp-pointed scissors or a scalpel. First remove strapping from the fine or delicate parts of the specimen then proceed to the more robust parts. If this is not done the heavier parts can spring up when released causing damage. Labels should be treated as above.

MIXED SHEETS

Whether or not mixed sheets should be separated is a vexed question only to be answered after careful thought; all such decisions should be taken by a botanist.

Mixed sheets should *not* be separated in the following situations:

- the specimens all belong to the same species (or infraspecific taxon) and are from the same area; separation would be unnecessary.

- the sheet is based on a mixed gathering where the material is intertwined and separation would cause damage.

- it is not entirely certain which specimens belong to which data labels and/or determination slips and paper capsules. Serious errors could result from separation.

- evidence of changing taxonomic opinion would be lost: e.g. a sheet of two gatherings has one determination slip by one botanist and two determination slips indicating different taxa by a later botanist. Presumably, the first botanist considered the two specimens to belong to the same taxon; this evidence would be lost by the separation of the sheet.

Two alternatives to separation can be considered:

- file mixed sheets in a separate cover, labelled 'mixed sheets', and place before the 'spp.' covers, or file in a 'mixed sheet' genus cover if different geographical areas are concerned. (See **16, Incorporation (filing or laying-in) of mounted specimens**).

- make a photocopy (or copies if more than two specimens) of the sheet and file as appropriate as cross-references. Indicate clearly on the photocopy which specimen is intended and where the original sheet is filed.

Great care must be taken when **photocopying specimens**. *Not all* specimens and *not all* photocopiers are suitable. Archival quality photocopy paper is available, but is more expensive than standard paper. If there are any loose items on the sheet they should be put in a paper capsule before the sheet is inverted onto the photocopier. The specimen must not be subject to crushing from the photocopier lid. If the copier has a photograph enhancer, the best results will be obtained when it is used.

If the sheet cannot be photocopied, cross-reference the specimens by means of dummy sheets (see **13. Ancillary collections**, *Recording and incorporation*) or photographs can be prepared (see **26. The photographic copying of herbarium sheets**).

Sometimes, however, separation of mixed gatherings is desirable and can easily be made.

- If possible, make photocopies (see above) of the complete sheet (one copy for each specimen). Check the back of the sheet for writing and photocopy if necessary. Fold the photocopies and place inside a large paper capsule or envelope, noting on the outside 'photocopy of original mixed sheet, separated on [date]'.

- If the sheet cannot be photocopied, cross-referencing labels must be prepared, they should state 'separated from . . . (collector's name and number, species name and country if necessary) on [date]'.

- Any detached items should be associated with a specimen only if it is *absolutely* certain they belong together; often this cannot be known for certain. Items of doubtful origin should be mounted with one of the sheets in a paper capsule, clearly annotated to show that the contents came from a mixed sheet; a cross-reference is necessary for the other sheets.

Then proceed as follows:

- cut around the specimens adopting the same procedure above, see *Re-backing and re-mounting*, keep as much of the original sheet as possible and leave the labels in their original positions if possible.

- place each specimen towards the centre of a new mounting sheet. A new label indicating the minimum data can be placed in the bottom right corner if desired, and paper capsules positioned as necessary.

- glue the specimen and/or labels and paper capsules in position, place under waxed paper and leave under a sandbag until dry.

- if the original glue has deteriorated, either introduce new glue under the loose parts of the specimen or totally mount afresh. This will need careful consideration (see above), but if done, ensure anything written or stamped directly on the sheet is cut out and mounted with the specimen.

- if the specimens were mounted by the strapping method, renew or re-glue the straps as necessary.

- any thick or heavy portions may need to be reinforced by stitching though the new mounting sheet (see **9. Mounting herbarium specimens**, *Sewing*).

REFERENCES

CCI: TB11 (1990)

11. PLANT NAMES (NOMENCLATURE)

PURPOSE

Scientific names of plants are a convenient shorthand way of referring to them and they form a vital key to the literature which contains the known information relating to plants. The names are thus an essential means of communicating information about plants. (see **1. What is taxonomy?**)

STRUCTURE

Names of plant taxa at all ranks are written in Latin or in Latinized form in accordance with the rules of the International Code of Botanical Nomenclature (ICBN). A *taxon* (plural *taxa*) is a taxonomic group of plants of any rank (e.g. the family *Rosaceae;* the genus *Rosa* and the species *Rosa canina* are all taxa). The name of a **species** consists of two words, and it is therefore known as a **binomial** (= two names), e.g. *Caesalpinia pulcherrima.* The first word is the name of the genus (**generic name**) and always begins with a capital letter. The second word is the **specific epithet** and always begins with a small letter.

Names of **families** are composed of the name of the type genus, with the ending (suffix) 'aceae' added. Thus, the name *Sterculiaceae* is based on the generic name *Sterculia.* However, alternative versions are permitted for the following long established family names:

- Compositae = Asteraceae, type *Aster*
- Cruciferae = Brassicaceae, type *Brassica*
- Gramineae = Poaceae, type *Poa*
- Guttiferae = Clusiaceae, type *Clusia*
- Labiatae = Lamiaceae, type *Lamium*
- Leguminosae = Fabaceae, type *Faba*, for the family in the broad sense. When divided into 3 families: Papilionaceae (or Fabaceae), Caesalpiniaceae and Mimosaceae.
- Palmae = Arecaceae, type *Areca*
- Umbelliferae = Apiaceae, type *Apium*

The names of **genera** are Latin or Latinized nouns, e.g. *Panicum, Elaeis, Adansonia.* These can be abbreviated to the initial letter when, for example, several species of the same genus are being referred to and there can be no confusion as to the genus concerned.

The **specific epithet** usually acts as an adjective and typically has a Latin ending. It can describe the species, e.g. *Caesalpinia pulcherrima* means the 'very beautiful Caesalpinia' and *Caesalpinia echinata* means the 'spiny Caesalpinia'; or it can indicate the geographical origin or ecology of the species, e.g. *Pavetta rwandensis* means the 'Pavetta from Rwanda', and *Pavetta saxicola* means the 'Pavetta that grows on rocks'. Also, people can be commemorated by specific epithets, e.g. *Rosa smithii* = the rose of Smith; *Rosa smithiana* = the Smithian rose.

When the specific epithet acts as an adjective it must agree with the genus in gender; epithets ending in '*a*' are usually feminine, '*us*' masculine and '*um*' neuter. Sometimes changes are necessary, e.g. *Canthium hispid**um*** becomes *Keetia hispid**a***, see below under '*double citation*'.

Sometimes taxa below the rank of species (or **infraspecific taxa**) are recognised, and must be included in the name. The most commonly encountered are subspecies (*subsp.* or *ssp.*) and variety (*var.*), but subvariety (*subvar.*) and forma (*f.*) are also occasionally used. Infraspecific taxa are ranked as follows:

subspecies

variety

subvariety

forma

Any one or more of these ranks may be used in a plant name. If more than one is used the correct sequence must be followed, e.g.

- *Alysicarpus glumaceus* subsp. *glumaceus* var. *intermedius*
- *Alysicarpus glumaceus* subsp. *macalusoi*
- *Alysicarpus vaginalis* var. *villosus*
- *Rosa stylosa* f. *variegata*

AUTHORS

The plant name when written in full is followed by one or more personal names; these are frequently abbreviated in form, e.g. *Ranunculaceae* Juss. (or Jussieu); *Commelina* L. (or Linnaeus); *Acacia schweinfurthii* Brenan & Exell (& or *et* = and). The correct abbreviations of authors' names are given in Brummitt & Powell (1992) (see **22. Essential herbarium literature**).

This personal name (or names) is known as the **author citation** for the plant name they follow. It is the name of the person (or persons) who first 'validly published' the name (see below, *rules*).

Occasionally '*ex*' connects two personal names, e.g. *Poinciana regia* Bojer *ex* Hooker. This indicates that the *first author* originally applied the name to the taxon but did not validly publish it, while the *second author* validly published the name, acknowledging the contribution of the first author. The *first* name may be omitted for the sake of brevity, e.g. *Poinciana regia* Hooker

Infrequently '*in*' connects two personal names, e.g. *Viburnum ternatum* Rehder *in* Sargent. Here the *first author* validly published the name but in a work otherwise written (or edited) by the *second author*. In this case the second name may be omitted, e.g. *Viburnum ternatum* Rehder.

Often two authors' names are cited but the first name is between brackets (parentheses), e.g. *Keetia hispida* (Benth.) Bridson. This indicates that the *first author* validly published a name for the taxon but at a different rank or in a different genus, i.e. *Canthium hispidum* Benth. – this is known as the **basionym**. The *second author* then revised the taxonomy of the taxon and gave it the present name or **combination**. This form of author citation is known as a **double citation** and indicates that there has been a change in taxonomic position or rank. (see *comb. nov.* in **23. Rearrangement of herbarium collections according to new publications**).

RULES (only a simplified version is given here)

All matters dealing with the application of names to plants are dealt with in detail in the International Code of Botanical Nomenclature (ICBN) and the International Code of Nomenclature of Cultivated Plants (ICNCP). A working knowledge of the *most recent* version of the Code is essential to a taxonomist.

A name must fulfil certain requirements of the ICBN if it is to be considered as the correct name for a plant (or taxon) (see fig. 21). It must be:

- *effectively published*: published in printed form and made available to the public or at least to botanical institutions.

- *validly published*: in an approved form and accompanied by a description or a reference to one (plus other requirements), see **23. Rearrangement of herbarium collections according to new publications**, *Invalid name.*

- *legitimate*: in accordance with all the rules. The commonest kind of illegitimate name is the *homonym*, see **23. Rearrangement of herbarium collections according to new publications**, *Illegitimate name.*

- *correct*: within any taxon the earliest published name, provided it is effective, valid and legitimate *must* be chosen, according to the **rule of priority**. All other names are **synonyms**, see **23. Rearrangement of herbarium collections according to new publications**, *Synonyms.*

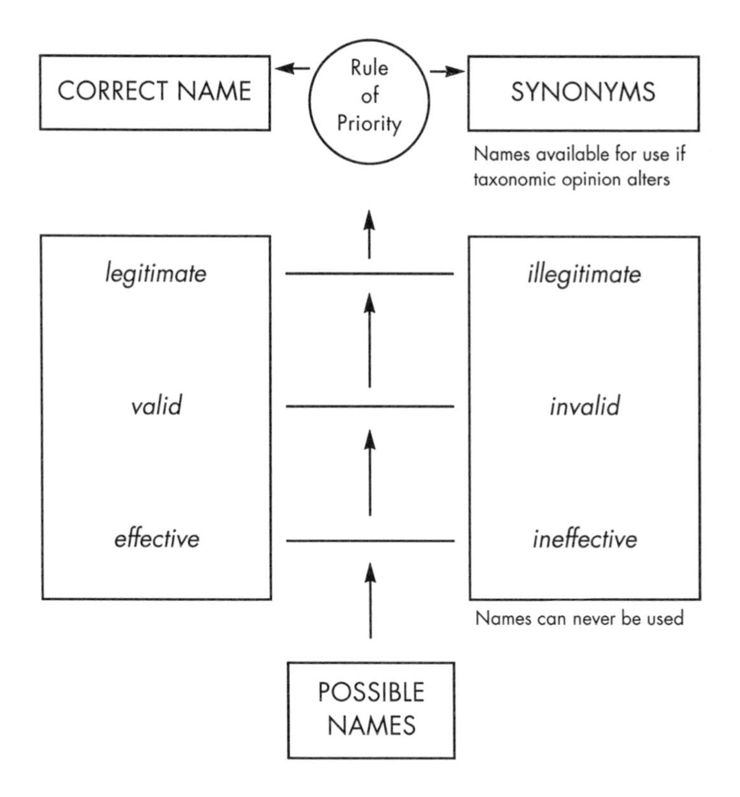

FIG. 21 — Stages in determining the correct name for a taxon.

TYPIFICATION (only a simplified version is given here)

Names are attached to taxa by means of **types**. At species level, for example, a type is *the* specimen (sometimes illustration) which formed the basis of the original description. Under the rules of the Code, the name and its type specimen are permanently linked. This means that if a type specimen is reclassified into another genus, the epithet originally attached to the type must also be transferred as in the above example of *Canthium–Keetia*. (The different kinds of types are described in **23. Rearrangement of herbarium collections according to new publications**).

NAMES OF HYBRIDS AND CULTIVARS

Names of hybrids are formed by placing × (hybrid sign) between the names of the parents, e.g. *Camellia japonica* × *C. saluenensis* or, alternatively, by use of a Latin collective epithet, e.g. *Camellia* × *williamsii* [= *C. japonica* × *C. saluenensis*]. However, if the hybrid is between species of different genera the × should be placed first, e.g. × *Fatshedera lizei* [= *Fatsia japonica* × *Hedera helix*].

The prefix *notho-* (or *n-*) is used to indicate a taxon (of any rank) known to be of hybrid origin, e.g. nothospecies, nothovar. etc.

Cultivar names follow the names of the species or hybrid concerned either preceded by 'cv.' or written within quotation marks, e.g. *Taxus baccata* cv. Variegata or *T. baccata* 'Variegata'. New cultivar names are not in Latin form, e.g. 'Loveliness'. All cultivar names are written with capital initial letters.

REFERENCES

Jain & Rao: 95–107 (1977)
Jeffrey (1973)
— (1982)
Stace: 211–217 (1989)
Stearn (1983)
Veldcamp (1987)

12. THE ARRANGEMENT OF HERBARIUM COLLECTIONS

The Herbarium is a storehouse of valuable information – but it must be arranged in such a way that this information can easily be retrieved. There are several ways of arranging a herbarium, and the choice of a method depends on many factors. How big is the collection, and how big will it eventually be? Who is to use it – multi-disciplinary workers or only specialist staff? Will the less skilled also have access? How many people are available to put it in order and maintain and revise this order? How much use will be made of the herbarium, and for what purposes will it be used?

Essentially, there are two basic ways of organizing a herbarium – the *alphabetical* and the *systematic*.

ALPHABETICAL ARRANGEMENT

Here the families are arranged alphabetically, as are the genera within each family and the species within each genus. It is, however, usual first to group ferns and fern allies, gymnosperms, and flowering plants into separate units, and also to divide the flowering plants into monocotyledons and dicotyledons, and then arrange these sub-groups alphabetically.

Advantages:

- Easy for the non-specialist to find taxa.
- Mounted material can be added to the herbarium by unskilled personnel.

Disadvantages:

- Related and therefore similar taxa are placed far apart, so that identification by matching becomes more difficult.
- Continual incorporation by unskilled personnel can lead to a build-up of errors, e.g. synonyms and spelling mistakes may not be noticed.
- Because an index is usually considered superfluous, a taxon can be filed under an old synonym and thus be represented in the alphabetical sequence more than once or only under the incorrect name. This is especially likely to happen at species level.

Note that while alphabetical arrangements are infrequently used in major herbaria at the family level, they are often used for genera and species.

SYSTEMATIC ARRANGEMENT

The arrangement of families

Here the families are arranged according to one of several **phylogenetic systems** which place supposedly closely related families together. Various systems are in use; the choice of a particular one usually depends very much on what the original organisers of the herbarium were familiar with. All systems have disadvantages and none is perfect. Once one has been adopted it usually remains in use for evermore, as the effort and time necessary for a change to another system effectively prevent such an alteration.

Advantages:

- Similar families are placed close to one another, so that identification by matching is easier.

Disadvantages:

- Can be difficult for the non-specialist to find families.

- Incorporation of material may require some specialist knowledge.

The systems in general use for flowering plants are:

- Dalla Torre & Harms, Genera Siphonogamarum (1900–1907). This is usually referred to as the Englerian system, as Dalla Torre & Harms' work is a numerical arrangement of the system of Engler & Prantl, Die Natürlichen Pflanzenfamilien.

- Bentham & Hooker, Genera Plantarum (1862–1883). This is the basis of the system used at Kew, and is also widely used in herbaria in Commonwealth countries.

Further works which may influence arrangements or modify them are:

- Hutchinson, The Families of Flowering Plants, ed. 1, (1926–1934). (Ed. 2 is not recommended).

- Cronquist, An Integrated System of Classification of Flowering Plants (1981). Again, a much more recent work, which has had some influence on thought about family recognition. A synopsis of this system is included in Mabberley (1987: 627–636).

- Dahlgren, Clifford & Yeo, The Families of the Monocotyledons: Structure, Evolution and Taxonomy (1985). This very valuable work makes many important changes to monocotyledon families, and some of its conclusions are becoming accepted, in particular the splitting of the old family Liliaceae into many smaller families.

All these works are either out-of-print or very expensive. A useful working alternative is Heywood (Ed.), Flowering Plants of the World (1978). This does not, of course, incorporate the conclusions of Dahlgren et al. A number of the smaller and more obscure families are omitted or included within larger ones. A summary of all major systems is included in Brummitt (1992); see p 129.

All these systems attempt to arrange families in a logical sequence, so that those closely related are filed close to one another. However, no linear sequence can accurately represent a multi-branched phylogenetic sequence and there will inevitably be large discontinuities. In most cases, both the main systems place the same families close to one another; they differ rather in the way in which the blocks of families are arranged.

The arrangement of genera within families

For identification by searching and matching it is clearly desirable to have the genera systematically arranged within families. However, there is no modern text that deals with the systematic organisation of all genera. If a modern systematic treatment for a complete family is available then it can be used. The following examples are recommended:

- Clayton, W.D. & Renvoize, S.A. (1986). Genera Graminum, Grasses of the World. Kew Bulletin Additional Series XIII.

- Kubitzki, K. (1990). The families and genera of vascular plants, 1. Pteridophytes and Gymnosperms. Springer-Verlag.

- Pennington, T.D. & Styles, B.T. (1975). A generic monograph of the Meliaceae. Blumea 22: 419–540.

- Uhl, N.W. & Dransfield, J. (1987). Genera Palmarum. A classification of palms based on the work of Harold E. Moore Jr. The L.H. Bailey Hortorium and The International Palm Society

In a local herbarium, if the home country is covered by a major flora then the arrangement used in it should be followed.

An index, if possible stating the arrangement used, should always be inserted at the beginning of large families.

The arrangement of species within genera

The problems here are similar to those of genera discussed above. As a general rule, if a local or regional flora is available then it should be followed, particularly if its arrangement is systematic. If there is no local or regional flora then other solutions must be sought. If there is a regional revision or world monograph, then it should be followed, so long as it is available in the Herbarium or its associated library. If there is nothing available then the personal ingenuity of the curator or technician can come into play, devising informal arrangements which place similar species together to aid identification by searching and matching.

As with generic arrangements, it is most important that the system in use within large genera is described, indexed and filed at the beginning of the genus. The author and reference to the arrangement should be clearly stated.

Geographical separation

Many herbaria have an additional geographical classification superimposed on the taxonomic arrangement. At Kew, for instance, the world is divided into 19 major regions, which are themselves subdivided into areas. This system is used to subdivide the collections – thus, at Kew, all the species of *Trifolium* from Europe (Area 1) are placed before those from the Middle East and North Africa (Area 2), and so on (see fig. 23, p. 108, for map).

Whether a herbarium adopts such a system depends largely on its size, the area it serves, and the diversity of this area. It facilitates naming within any one area, but wide-ranging (and often variable) species will be dispersed in the geographical sequence.

Cultivated plants

These are often filed separately at the end of the other material of the species, genus, or family, sometimes in distinctively coloured folders. In a large herbarium this segregation of cultivated material facilitates the identification of cultivated specimens, especially if of unknown origin.

Ancillary collections

There are several kinds of material, apart from specimens mounted on sheets, which can usefully be accumulated by a herbarium. These include collections of bulky fruits and seeds (carpological collections), spirit (wet) collections of delicate or fleshy flowers, bulky or whole plants (e.g. cacti), illustrations and photographs. These and their arrangements will all be dealt with in more detail in **13. Ancillary collections** and **14. Collections of illustrations and photographs**. The importance of cross-referencing between ancillary collections and the main collection cannot be over-emphasized.

Special and subsidiary collections

These are collections of herbarium sheets which are kept separate from the main herbarium for one reason or another.

– *Historic collections* are often separated for their own safety as they are irreplaceable, usually fragile, and important to specialists but of limited use to the casual user or for identification. Historical collections should be arranged in their original sequence. Attempts to update the nomenclature should be avoided except by a cross-referenced catalogue. At Kew, the *Wallich Herbarium* is an example.

– *Types* are sometimes kept completely separate from the rest of the Herbarium for safety and ease of removal in a crisis, and to protect them from regular handling. In many herbaria they are filed in the general sequence but in special coloured or red-edged covers for protection and ease of location.

– *Reference collections for routine identification* can sometimes usefully be kept separate from the main herbarium. These should comprise 1–3 good and representative sheets of each of the more common species of the region. They will allow most routine identification to be carried out without using the main herbarium, and would be particularly useful in, for instance, University Botany Departments where students are required to make collections and identify them. Their use avoids excessive wear and tear on the main herbarium. The specimens in routine identification collections should be placed inside protective transparent sleeves.

Whatever the kind of collection and its manner of arrangement, it is important that the cabinet doors, shelves, boxes or drawers are clearly and unambiguously labelled and that the sequence from one block of furniture to the next is logical and easily understood by staff and visitors. In some herbaria a shelf marker, bearing the name of the family on a short overhang, is inserted at the beginning of each family.

The arrangement of fungi is covered in **17. Curation of special groups**.

REFERENCES

Brummitt (1992)
Bentham & Hooker (1862–1883)
Cronquist (1981)
Dahlgren et al. (1985)
Dalla Torre & Harms (1900–1907)
Engler & Prantl (1887–1915)
Heywood (1978)
Hutchinson (1926–1934)
Jain & Rao: 60–61 (1977)
Mabberley (1987)

13. ANCILLARY COLLECTIONS

SPIRIT COLLECTION

Although herbarium specimens are an ideal way of preserving the majority of plants, there are some groups of plants, such as orchids and succulents, which make poor specimens when pressed. In such cases it is often appropriate to preserve material in 'spirit', a preservative liquid such as Copenhagen mixture (see below). Representative flowers and soft fruit from non-succulent plants are also usefully preserved in spirit. Inflorescences of Compositae, for example, become distorted and fan-like when pressed and the arrangement of the involucre is lost, and the exact aspect of the corolla segments in a Papilionaceous flower can become obscure. Since specimens preserved in spirit retain their life-like appearance, botanical illustrators should work from spirit material whenever possible. Measurements taken from spirit material (especially the more fleshy organs) are usually more accurate than those based on dry material where shrinkage may have occurred. (See also **17. Curation of special groups**.)

Preservatives

Before storing in a preservative mixture, specimens must be adequately *fixed* in a suitable *fixative* when collected – this is to minimise distortion of the tissues (see **31. Collecting materials for ancillary disciplines**, pp. 227 & 230). After fixation the material is then transferred to a preservative mixture. The preservative mixture used at Kew is Copenhagen Mixture: 70% alcohol (ethanol or IMS), 29% water and 1% glycerol. The glycerol prevents the material from becoming too brittle. Particular care must be taken with alcohol because it is highly flammable. If possible a fume-cupboard should be used, otherwise it must be used in a well-ventilated place to prevent inhalation of its fumes.

Housing and arrangement

Ideally a well-ventilated room should be set aside to house the spirit collection. If this is not available, the spirit material can be kept in a cabinet near the herbarium collections for convenience. Fire precautions must be observed (see p. 13).

Spirit material is best stored in glass jars. These should be in a number of different sizes, e.g. 70 cc. to 3000 cc., and have wide necks. Jars with ground glass stoppers are recommended. The ground glass surfaces must be lightly coated with petroleum jelly to prevent sticking and also to provide a seal. A type of jar with plastic polypropylene lids may be used but the lids do not last indefinitely. Metal lids should be avoided because corrosion results from contact with the preservative.

The jars are best arranged numerically with blocks of consecutive numbers allocated for each size of jar, and jars of any one size being stored together. New specimens are given the next available number in the appropriate size of jar without any attempt at systematic arrangement. It may, however, be advantageous to separate families such as Orchidaceae which have large numbers of spirit specimens. The jars can be stored on shelves in metal cupboards or, if not too large, in metal storage cabinets with drawers having an inner width of c. 22 cm.; the top drawer should be well below eye-level.

Recording and incorporation

- All spirit material must be cross-referenced, preferably including the jar-number, with the corresponding herbarium sheet. (see **8. Processing unmounted specimens**). If the specimen is 'spirit only' place a dummy sheet bearing the data and cross-reference in the herbarium.

- Record on a label in spirit-proof permanent ink (or use a typewriter):
  a. The scientific name (authority optional).
  b. Country of origin (or cultivation) and locality.
  c. Collector's name and number.
  d. Date of collection.
  e. Full colour notes and any other descriptive data.
  f. The number of the jar.

- Then place this label inside the jar (see p. 46, '*Resistall*'); a second data label can be stuck to the outside of the jar and varnished for protection or it may be attached to the neck of the jar.

- Write the jar number on the lid of each jar. The kind of pen for use on overhead projector acetate film can be used, but the number should be varnished for protection. Alternatively, an oil-based paint can be applied with a fine brush; small tins such as used by model makers are the best.

- Copy the above data on an index card and file it, or incorporate into a computer database.

- File the jar in the appropriate space.

Study of spirit material

Staff removing jars of spirit material for study should be asked to record the details in a **log book** (i.e. plant name, jar number, date of removal and date of return to the collection).

If the material has been stored in a fixative rather than a storage solution it should be rinsed in dilute spirit (50% ethanol) to remove all traces of formalin before examination.

Maintenance

The collection should be checked for evaporation regularly and extra preservative mixture added where necessary. If evaporation is over 10% the mixture should be completely replaced.

Loans

If specimens are sent on loan, the partial or total contents of each jar should be transferred to leak-proof plastic bottles for transit and a record of the loan entered in a log book (see **19. Loans to other institutions**). The bottles should be topped up to the rim as bubbles of air can damage plant material.

CARPOLOGICAL COLLECTION

Dried specimens, mainly fruits but also portions of stem, bark, inflorescence or roots and tubers etc., which are too large to mount on herbarium sheets are stored separately together with a data label. Carpological material must be cross-referenced with the corresponding herbarium sheet or, if there is carpological material only, a dummy sheet bearing the data and cross-reference must be put in the herbarium.

The specimens are placed in boxes of various sizes with tight-fitting lids. If possible the lids should be topped with glass or a thick transparent polyester film so that the contents of the box are visible. A small label indicating the family, scientific name, collector and collector's number should be firmly attached to the lid.

The boxes are conveniently stored on trays on shelves, or in shallow drawers. Alternatively, the specimens can be placed inside individual polythene bags, labelled and stored in the type of boxes used for palms (see **17. Curation of special groups**). This method is not suitable for prickly or fragile specimens, nor for use in very humid conditions.

Carpological collections should be housed separately but ideally kept as close to each family as practical. The sequence of families and genera should be the same as in the herbarium, but for simplicity the species can be arranged alphabetically without geographical separation.

Some fruits (e.g. *Sterculia*, *Sloanea*, *Baphia* and many conifers) although apparently dry can dehisce and release their seeds. Such fruits can be tied with string or secured inside mesh tubes to try and preserve their original shape and avoid loss of seeds.

If the boxes are provided with tight-fitting lids which can keep out insects, it is not necessary to disinfest carpological specimens. However, in the tropics regular inspection is recommended and an insect-repellant added if necessary. See Jain & Rao 1977.

WOOD COLLECTION

If the herbarium has no wood collection, small portions of stem or trunk can be kept with the carpological collection.

Wood samples can either be arranged by accession number or systematically within a series of given size ranges. It is not necessary to place the samples inside individual containers. The smaller specimens can be housed in drawers and the larger ones in cabinets or on shelves. It is very important that the data label is securely attached to the specimen; if the sample has been prepared to show the timber qualities the data can be glued directly to it (in humid tropics water-based glues should be avoided). If the specimen still has bark present, tie on the data label or secure it in some way. Sometimes it may be possible to mark the accession number directly on the specimen with steel dies. If the bark has separated from the wood, it should be securely tied in position.

The specimens should be cross-referenced with their herbarium sheets and the relevant data copied on reference cards or a computer database in the same way as the spirit collection.

As new wood samples arrive they should be placed in the deep-freezer for a week in order to kill any insect or other infestation.

SEED COLLECTION

For the identification of seeds it is useful to build up a seed collection. Glass tubes with polythene stoppers are very useful for the storage of small seeds. Tubes 5 cm. long by c. 12 mm. and 15 mm. diameter are suitable sizes. The essential data are entered on a small label inserted in the tube so that the information can be read through the glass but also allowing the seeds to be seen. Cross-references to voucher herbarium specimens must be made as necessary.

Larger seeds can be stored in transparent-topped boxes or, if the conditions are sufficiently dry, in polythene bags.

Herbaria which receive from coastal locations **drift seeds** for identification should build up a special collection of these seeds for comparison.

REFERENCES

Jain & Rao: 71 (1977)
Stern in Fosberg & Sachet: 63–64 (1965)
Womersley: 95–98 (1981)

14. COLLECTIONS OF ILLUSTRATIONS AND PHOTOGRAPHS

GENERAL ILLUSTRATION COLLECTION

A collection of illustrations is an important supplement to the herbarium collections by providing information on, for example, habit and flower colour. Illustrations are especially important for succulent plants, and they are often helpful for identifying cultivated plants of unknown origin.

The **sources** of illustrations are varied; some examples are the following:

— Watercolour paintings: these may be commissioned, purchased or presented. Special care should be taken of them, as good examples can be valuable.

— Black and white scientific illustrations: these are usually commissioned to accompany botanical publications.

— Photographic prints of living plants: general habit views and close-ups of flowers or fruit.

— Colour prints: these may be acquired from various sources such as page-proofs, or accurately named pictures on calendars and post-cards.

— Illustrations cut from magazines or newspapers. If possible, substitute a photocopy as the paper will last longer and not spoil the surrounding sheets (as would highly acidic newspaper).

Housing and arrangement: In order to standardise the collection, the illustrations should be mounted on sheets of the same size as the herbarium mounting sheet and kept in cabinets near the herbarium collections.

Any large sheets will need to be housed separately, but then a dummy sheet should be placed in the standard collection with a cross-reference.

Valuable paintings should be kept separately in the library under archival and secure conditions. If possible store in made-to-measure conservation-quality boxes.

The illustration collection is most usefully arranged by family and genus in the same arrangement as the herbarium, but species are best arranged alphabetically, without geographical separation. The genera should be grouped together in labelled folders. If the illustrations are to be filed in this manner it is not necessary to keep an index to species. However, synonyms must be cross-referenced and if the illustrations are valuable an index can help save unnecessary handling.

Accessioning: All valuable items should be accessioned and a record kept. If possible mark the margin with the institution's embossing stamp to deter theft.

An index to artists with a list of species drawn by them may be worth compiling.

Mounting illustrations

Careful and correct mounting of illustrations is essential if they are to last as useful reference material.

Ideally each illustration should have a backing mount *and* an overmount or window mount (see fig. 22). Both of these boards should be good quality mount boards, either Conservation board or Museum board (more expensive).

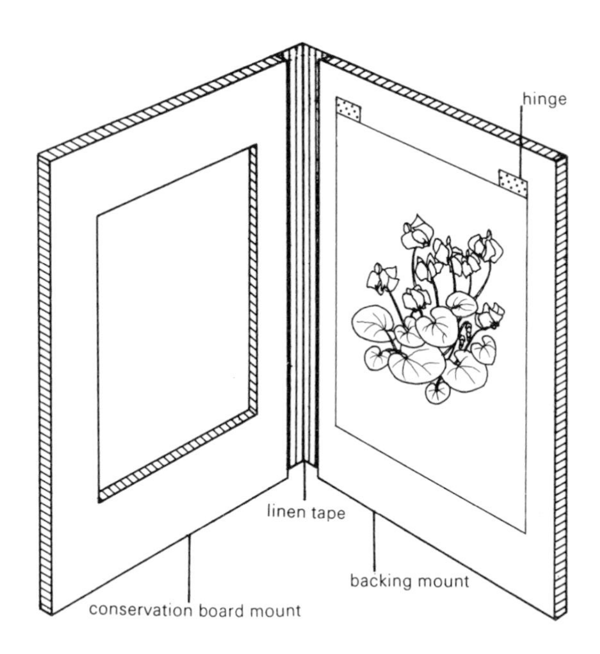

FIG. 22 — Window mount.

Unfortunately cost and lack of storage space often means that a thinner good quality backing mount is all that can be provided, e.g. a 300 gsm. chemical wood pulp, archivally sound paper. In this event illustrations should be interleaved with an acid-free tissue or enclosed in polyester sleeves to protect the surface. (Polyester sleeves can be problematic in humid climates if condensation forms between illustration surface and the sleeve). For practical reasons herbarium mounting sheets can be substituted (see **5. Materials**).

Illustrations should remain as free as possible in order to allow them to move in response to change in temperature and relative humidity. They should *never* be secured at the corners or edges to the backing mount, or in any way to the overmount, or splitting and buckling could occur. Instead hinges are used: there are three types (see fig. 23).

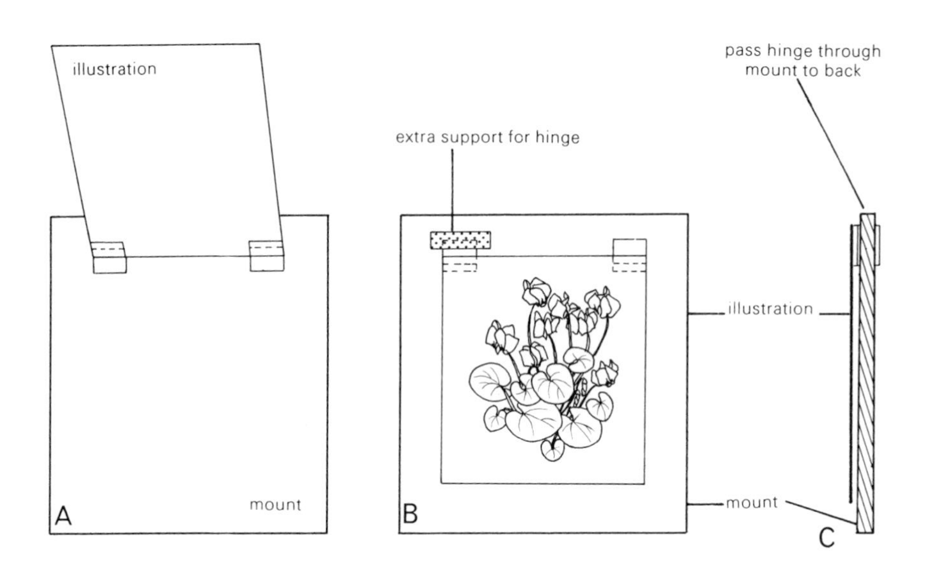

FIG. 23 — Types of hinge. **A** fold-under hinge; **B** drop hinge; **C** through-mount hinge.

Normally illustrations are only hinged at the top – however, there are occasions where hinges at the bottom will give added protection, e.g. where there is no overmount, and researchers are not trained in correct handling of mounted illustrations.

Materials (see **5. Materials**)

Hinges: Acid-free paper, *lighter* than the illustration paper:

- For lighter works – Japanese tissues/papers, lens tissue, 'M tissue'.
- For heavier works – heavier Japanese papers or Western paper of conservation quality. White air mail paper can be substituted.

Adhesives for hinges: Must be reversible and non-staining e.g.:

- Starch paste (10% weight/volume). See CCI, 1990: N11/4.
- Methyl cellulose.
- Some commercial starch paste preparations may be acceptable and have the advantage of being thicker and faster drying than homemade wheatstarch paste.

It is not advisable to use PVA glues as they are not easily reversible (i.e. water soluble). 'Sellotape' and other self-adhesive tapes must *not* be used.

A data label should be glued to the bottom right corner of the mounting board, including the following information:

- Name of plant.
- If a photograph – photographer, locality, date, and if a specimen was collected, the collector's name and number and a cross-reference.
- If a published drawing or watercolour – the artist and reference to the publication.

PHOTOGRAPHIC RECORDS OF HERBARIUM SPECIMENS

These may be required for the following reasons:

- Records of type or other historical specimens received on loan.
- Records of specimens received on loan of new taxa or taxa not represented in the collections.
- Records of type specimens to be sent out on loan to other institutions.

Photographic records are usually either black and white prints, Cibachrome prints or microfilm (see **26. Photographic copying of herbarium sheets**). See CCI 1990: N16/1–N16/6; DePew 1991: 195–207.

Photographic prints can be mounted on herbarium sheets and then filed into the collections along with the specimens. They will, however, need protection since, apart from the risk of the print getting scratched by woody or spiny herbarium specimens there is the risk of deterioration caused by corrosive substances from the insecticides used on herbarium specimens. Mount either by:

- Glueing an open-topped envelope to the centre of the mounting sheet and placing the print inside.

- Hinging at all four corners or, if something is written on the back, the print may be hinged at the top and supported by photographic corners at the lower corners. Protect the print from the surrounding herbarium sheets either by covering with a top mount or placing inside a polyester sleeve or a species (or type) cover of its own.

Attach a label to the bottom right-hand corner giving the following data:

- Name of plant.
- If a type, then name of which it is a type and type status.
- Locality.
- Collector's name and number, and date.

Microfilm: best stored by cutting into either:

- Lengths and placing inside transparent sleeves. A sleeve taking two lengths of film with a frosted upper margin on which to write the data is commercially available.

- Individual frames and placing inside aperture cards, i.e. a transparent pocket set in a card on which the data are written.

Both kinds of display sleeve or card can be filed in taxonomic order, so no index is necessary.

Some form of viewing apparatus (similar to microfiche viewers) should be kept near the file or microfilms.

35 mm. TRANSPARENCY COLLECTION

The main uses of 35 mm. transparencies are to provide visual aids for lecturing, and to supplement information present on the herbarium sheet, although it is also useful to obtain a print from the transparency and file it in the general illustrations collection.

Accessioning and indexing: The simplest way is to mark and file each transparency according to an accession number, make out an index card indicating the subject matter and accession number, and file it in subject order, or enter these data on a computer database.

If the collection is large it may be advantageous to divide it into separate smaller units according to broad subject headings, e.g. habitat and ecological collection, historical collection, as well as a systematically arranged botanical series.

Mounting, storage and preservation

Mounting – there are basically 2 types of mounts: mounts with glass and glassless mounts.

Original 35 mm. transparencies should ideally be mounted into *glassless* mounts.

Duplicates can be mounted into either mount but the glass mount provides greater protection against damage during handling and projection. However, glass mounts should be avoided in all areas of high relative humidity.

Storage – the valuable originals should be stored in their original boxes or in a suspended file system, ideally in total darkness, low temperature, relative humidity not higher than 30% and well away from any air pollutants. Ideally, they should never be used for projection or publication purposes.

Duplicates can be stored in any system most convenient for retrieval, not forgetting that they will also be adversely affected by careless handling, frequent and prolonged projection, high humidity and temperature, and excessive air pollution.

Preservation – all colour transparencies are adversely affected by excessive and prolonged exposure to light, high levels of temperature and humidity during storage, and exposure to atmosphere pollutants, certain gases and some reactive materials.

Certain types of colour film have greater image stability than others.

The Eastman Kodak Company publishes comprehensive literature related to the handling, storage and preservation of colour images.

REFERENCES

CCI: N11/4 & N16/1–N16/6 (1990)
DePew: 195–207 (1991)
Womersley: 100–104 (1981)

15. HANDLING HERBARIUM SPECIMENS

It is important to treat the herbarium as a valuable archive which must be preserved for future generations. Labels and annotations must remain legible for centuries. It is vital that specimens are always handled in a manner that will not cause damage or the loss of fragments. Many specimens are irreplaceable.

Good practice and hints

- Specimens should always be handled with great care.
- Hold specimens by both sides rather than the base; this is essential if the mounting paper is thin. If supported only at the base the paper can bend downwards and damage the specimen.
- Keep the sheets flat and fully supported, especially if carrying them from one place to another.
- Never shuffle specimens so that the sharp edges of the sheet cut the underlying specimen.
- Never align the specimens in a cover by holding it upright and striking one edge of the cover on the table; fruits etc. can become detached and fall out if this is done.
- Never turn the specimens as though they were the pages of a book (any loose fragments fall off if the sheets are placed face down). If you wish to look through a folder, put it flat on the table and stack the specimens neatly to one side, always keeping them horizontal.
- Before returning a cover to the cabinet check that all the sheets are aligned so that none protrude and are liable to be bent.
- Never bend a specimen to examine part of it under a stage microscope. Use a long-armed microscope if possible, or hand-lens.
- Never rest books, heavy objects or elbows on unprotected specimens.
- Place loose fragments in a paper capsule and fasten it to the sheet, providing it is certain that they belong. Always use the approved adhesive, never a convenience product such as 'PrittStick'.
- When genus covers are placed in the cabinets, the open edges should contact the right hand side of the pigeon-hole. This will give access to the spines of the covers (if the cabinet is sufficiently wide). If the open edge is carelessly handled, the fingers can be cut by the mounting sheets.

- Take care to replace both genus and species covers in the correct sequence.

- When removing a genus cover from the cabinet, the cover immediately beneath the cover being removed should be pulled forward a fraction to indicate the correct position. Much time can be wasted if you have to search for the position when you replace the cover.

- Do not force too many specimens into a shelf or storage box.

- Always make sure cabinet doors are firmly shut. Dust and insects can enter if doors are left ajar.

- Before removing any specimen from the herbarium (e.g. for loan) check that any essential data present on the species cover are noted on the sheet (add in pencil if necessary). Older sheets often do not state the country and sometimes the specific or infraspecific names have not been updated.

- If any specimens or folders are removed from the herbarium for study elsewhere in the building, e.g. for artists or botanists working in offices or separate rooms, insert a hanging label to indicate taxa/taxon removed, present whereabouts and date of removal.

- Any specimens in need of repair, reglueing or possible separation (older sheets often consist of more than one collection) should be dealt with as soon as noted (see **10. Conservation of sheets**). Insert a hanging label (as above).

- Never keep specimens out of the cabinets for longer than necessary. While on the bench they can be subject to damage, e.g. have heavy objects placed on top of them or be blown on the floor or get wet if near an open window. Sunlight can cause deterioration and specimens should not be exposed to it.

- If you wish to leave any specimens spread out in order, or in stacks, for a short period, first cover them with mill-boards (or cardboards) for protection. Unmounted specimens are especially vulnerable if left in a breeze from an open window. The stacks should never be so high as to be insecure.

- When sorting specimens with the data labels attached to the bottom of the sheets, arrange them in small overlapping heaps so as to keep the labels visible.

- Always keep a look out for insect damage. Any live insects or damage of possible recent origin *must* be reported immediately.

- Old determination labels must never be removed. New ones must be clearly written with the date and botanist's name in full; they should be attached with the approved adhesive.

REFERENCES

Fosberg & Sachet: 77 (1965)
Jain & Rao: 63–66 (1977)
Radford et al.: 763–764 (1974)
Steenis, van (1950)
Womersley (1981)

16. INCORPORATION (FILING OR LAYING-IN) OF MOUNTED SPECIMENS

NEW ACCESSIONS

The task of incorporating mounted specimens is very much easier in herbaria with alphabetical arrangements than in systematically arranged herbaria (see **12. The arrangement of herbarium collections**). Also, local herbaria usually do not have the complication of imposing a geographical sequence on their specimen order (see fig. 24). In a general herbarium dealing with plants from all over the world, the system becomes quite complicated and staff must be adequately trained to use the system. In a large international herbarium, specimens must be sorted in turn by family, genus, geographical area, species, and where applicable, infraspecific taxa and geographical location.

Whatever the arrangement, it is essential that sheets are filed into the herbarium in the correct place, otherwise they may be lost for a considerable period. If a specimen seems to be the first example of a species to be filed into a particular folder, one should be suspicious and check again before filing it. If there is any uncertainty about a locality, a comprehensive gazetteer or atlas must be consulted. If any doubt arises when filing a specimen, advice must be sought.

Specimens are protected and taxonomically grouped by placing them inside folders (ideally both genus and species covers) and any type specimens in type covers (see **5. Materials**). Some herbaria place each sheet in an **individual folder**, a procedure which is recommended for historical herbaria, but in an actively growing herbarium this involves considerable extra expense, labour and space; it also makes difficult the rapid consultation of specimens when naming.

The respective folders must be clearly marked with the necessary information:

- *genus covers*: family name (and number optional), generic name (author unnecessary) and number (if any), species epithets and numbers (if any), or initial letters, e.g. H–K, if alphabetical arrangement.
- *species covers:* country (or district), species epithet (and number), any infraspecific taxa.
- *type covers*: at least the name of which the specimen is the type; if it is a synonym then the currently accepted name should also be shown.

The exact manner in which the different covers are written up will vary in different herbaria, but the practice followed at Kew is illustrated (see fig. 25) and this can be adapted as appropriate for other institutions.

Species and genus covers must not be over-filled. Ideally there should be no more than about 3–4 species covers per genus cover, but obviously this varies with the thickness of the material. Bulky specimens can sometimes be turned around to even out thickness in a cover. Any specimen with a projecting part that

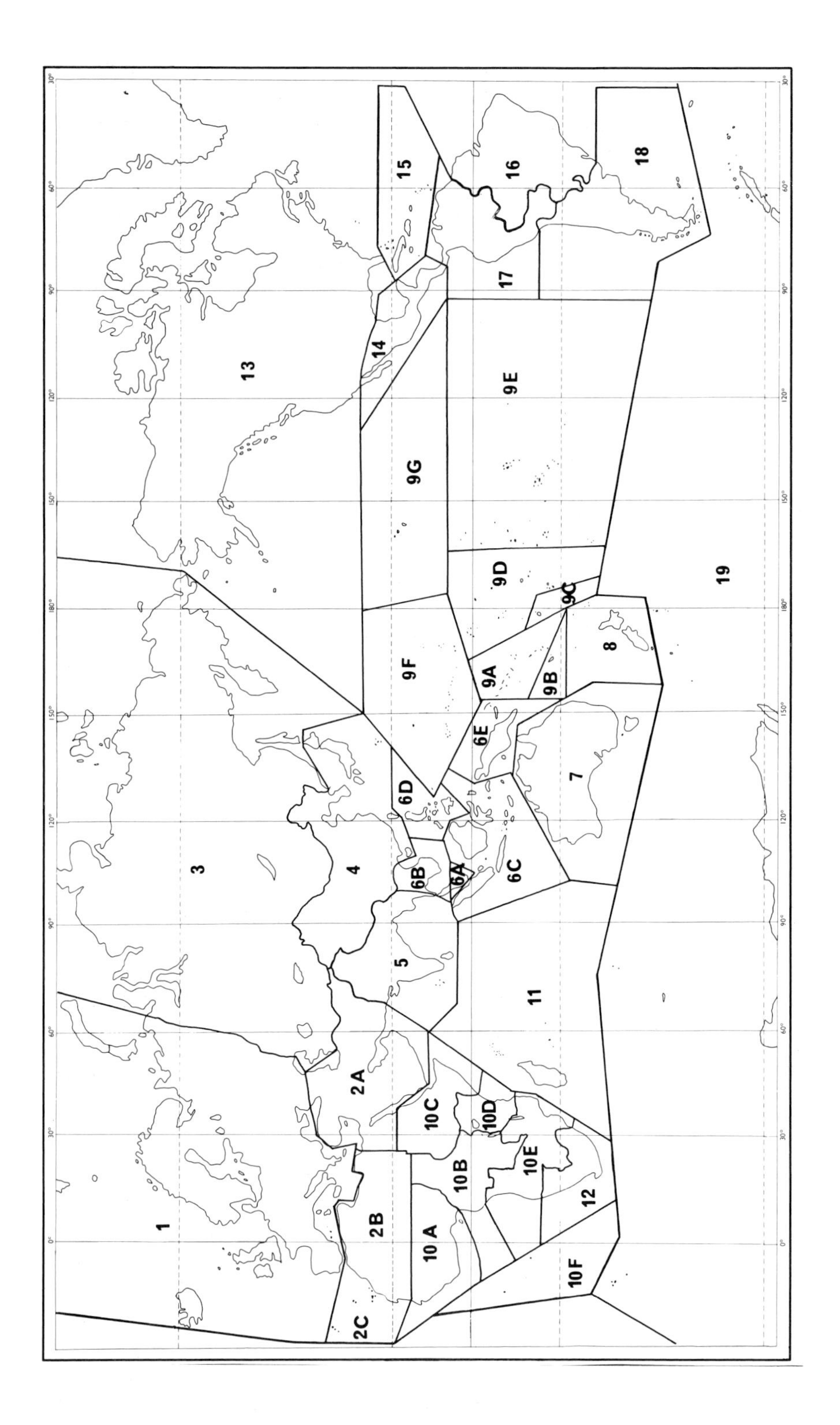

FIG. 24 — Example of geographical divisions of the world.

could damage the sheet placed on top of it should be put at the front (or top) of the cover and padded if necessary. Fragile or old and delicate specimens can be protected with individual covers or by transparent sleeves or type covers (see **10. Conservation of sheets**).

Damaged, fragile or dirty genus or species covers should be replaced as necessary. However, if they bear any labels or notes, other than the standard labelling to show taxon and locality they *must* be retained inside the new covers and not discarded.

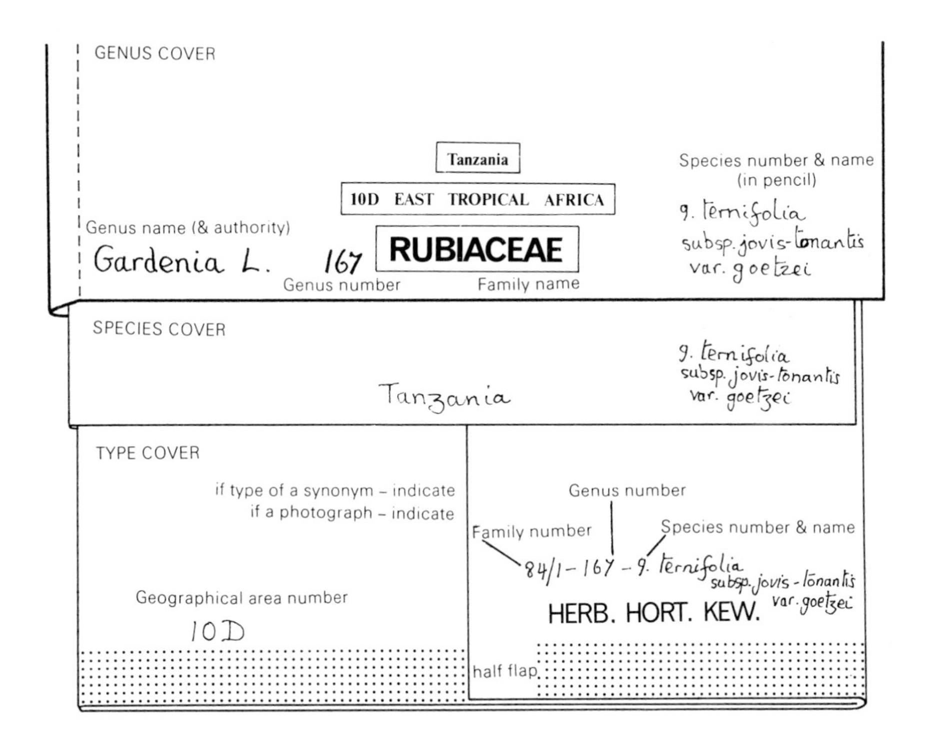

FIG. 25 — Genus, species and type covers written up with relevant data. Note: Kew genus covers do not show the family number.

Procedure for filing mounted specimens

The mounted specimens are sorted by the following categories:

a. *Family.* If not indicated on the label, generic names can be referred to family by using Brummitt, 'Vascular Plant Families and Genera', Mabberley, 'The Plant Book', or Willis, 'Dictionary of the Flowering Plants and Ferns'.

b. *Genus.* Generic indexes (indicated on the appropriate cabinet doors) give the genus numbers (if arrangement is systematic).

c. *Geographical area* (if applicable).

d. *Species.* Species may be arranged either in alphabetical or in systematic order according to a revision or flora, in which case an index should be provided wherever possible. If new species have been described after publication of the revision on which a numbered arrangement was based, add them alphabetically at the end of the numbered sequence. Alternatively, if a new species is very clearly related to a certain previous species, it can be added in as, e.g. '15.a' or '15/1' and the name entered in the species index.

Subspecies and *varieties* (and rarely *forms*) should follow the typical subsp. or var. of the relevant species. In species where subspecies and/or varieties are recognized, some specimens may be named only to the species or to the species in the broad sense i.e. '*sens. lat.*' (or '*s.l.*'). These specimens should be filed in their own species covers, separate from specimens named to subspecies and/or varieties.

Any specimens that are doubtfully named (i.e. have '?', 'cf.', 'aff.' or 'sp. near' before the specific epithet) should be placed after those of the same epithet that are definitely named.

Specimens named only to genus should be kept in a separate folder marked '*spp.*' at the end of the sequence. It should be noted that '*sp.*' is singular and '*spp.*' is plural for species. However, if a determination such as '*sp. A*', '*sp. B*' etc. or '*sp. 1*', '*sp. 2*' etc. has been used, this indicates that either:

– the specimen belongs to a published taxon lacking a formal name; probably there was too little material on which to base a formal description. Many floras and revisions include such taxa in their appropriate systematic order and this should be followed in the herbarium. Such taxa should be included at the end of the index to species. It is good practice to give some indication of the publication on the folder (e.g. '39. *P. sp. D* of F.T.E.A.' – i.e. *Pavetta* sp. *D* of the Flora of Tropical East Africa, Rubiaceae, part 2) so that confusion with '*spp.*' is less likely.

– some preliminary sorting of the indet. material (i.e. indetermined or unidentified) has been done prior to (or pending) revisionary work. The designated taxa could either be named species not represented in the herbarium or new species, it is not known which; or the correct name has yet to be decided because the taxonomy or nomenclature needs sorting out. An additional way of indicating this kind of informal determination is 'sp. = *Stringfellow* 123', 'sp. = *Ropeman* 999' etc. Such taxa should be grouped in separate folders labelled '*sp. 1*' etc., '*sp. B*' etc., or '*sp.* = *Stringfellow* 123' etc. as appropriate.

e. The *countries* within a geographical area should be arranged in a definite sequence which is followed for each species.

If a species within a single geographical area occupies more than one genus cover, the genus covers are given running numbers in pencil.

The recommendations for filing *exotic taxa* vary in different herbaria. *Introduced weeds* and *naturalized species* are often included with the main sequence. *Garden escapes* can cause confusion as they may not have been noted as such by the collector and are often not included in floristic or monographic accounts. *Cultivated* taxa are generally easier to note and deal with. If the collections are geographically sequenced, the cultivated covers can be placed at the end and clearly marked as such.

RETURNED LOANS

These are reincorporated in the same way as new accessions. If there are any redeterminations or unpublished names ('*ined.*') ask a botanist before attempting to file the specimen away. Not all revisionary work is reliable and some (e.g. student work) may never be published. The botanist may well advise filing the specimens back under their old names and then fully re-arranging the genus when the publication eventually appears; the botanist may even prefer to deal with returned loans personally.

REFERENCES

Brummitt (1992)
Mabberley (1987)
Mori et al.: 36–38 (1989)
Willis (1973)

17. CURATION OF SPECIAL GROUPS

Some groups of plants need special treatment in their processing and storing. For the curation of Bryophytes and Algae see chapters 33 and 34 respectively.

Succulents

Succulent plants (or succulent fruits or delicate or complex flowers) are best preserved in a spirit collection (see **13. Ancillary collections**). All specimens must be cross-referenced with the herbarium; if necessary a dummy sheet must be added in the herbarium showing the relevant data. Wherever possible succulent material should be complemented by colour photographs or drawings; illustrations must also be cross-referenced (see **14. Collections of illustrations and photographs**). N.B.: Flower colour is entirely lost in spirit.

Cactaceae can be dried (whole or halved) and stored in boxes with the carpological material (see **13. Ancillary collections**).

Palmae, Pandanaceae and Cyclanthaceae

Owing to the bulky nature of specimens of these families special problems arise. The need for maximum storage coupled with minimum damage and ease of access are harder to achieve than in the general herbarium collections. There are three main systems in use:

- ***The box and folder system*** is by far the best system, but it is the most expensive. *Palm-folders* should be made of archival quality card and have four flaps (see fig. 26). The folder, when closed, should have the same area as a standard herbarium sheet. The flaps are scored at the base with three lines at least 1 cm. apart so that the depth can be varied to fit the specimen. An indication of the scientific name of the specimen, collector's name and number and country of origin must be clearly shown on the outside of the right hand flap near the lower margin. The data may also be written on the front vertical edge of the folder so that they are visible when the specimens are filed. The original data label must be placed inside the folder together with the unmounted specimen. The folders may then be filed in the cabinets, directly on shelves or placed inside drop-front herbarium boxes (see **3. The herbarium building and specimen storage**, *Boxes)*, taking care to avoid crushing the specimens.

  Many palm specimens are too bulky for such folders and require storage in *palm boxes.* The boxes should have basal dimensions equal to those of the general herbarium drop-front boxes; they should have a top fitting lid but *no* drop-front. Ideally the boxes should be of two or three different depths to allow efficient storage of different sized specimens; the depths should represent fractions of the depth of the general boxes, e.g. those used at Kew are 6.6 cm., 13.3 cm. and 20 cm. deep. Unmounted specimens with their data labels, should be stored singly in such boxes and the lid and front labelled with the relevant specimen data. The palm boxes are easily interpolated into the general sequence, where the folders are stored in the drop-front boxes of equal basal dimensions.

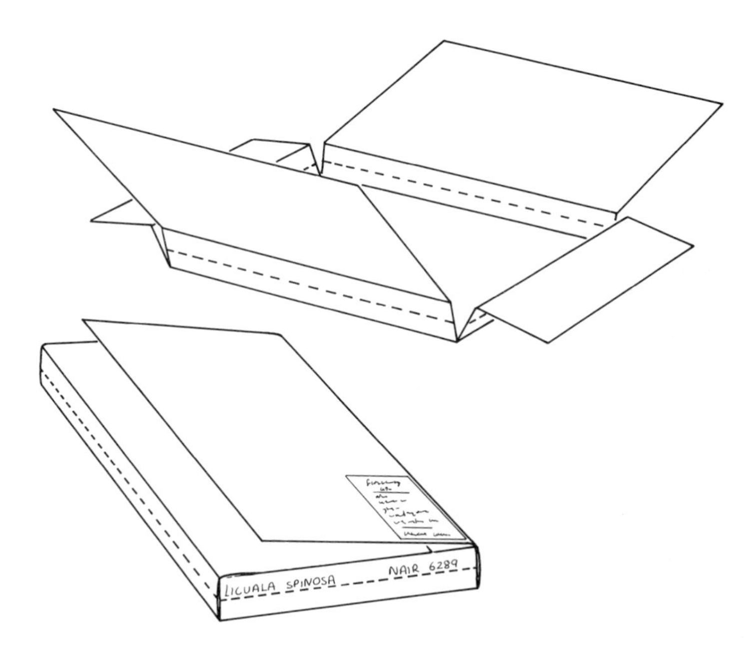

FIG.26 — Palm-folder; dotted lines indicate scoring to give choice of depths.

- ***The 'bag' method*** is suitable for dry climates. The specimens are placed unmounted on boards (e.g. corrugated) which are the same size as standard herbarium sheets. Labels with data are then secured with paper-clips to the board. Loose fruits and flowers are placed in separate polythene bags and laid on top of the main material; male and female flowers are kept in separate polythene bags. The whole specimen is then enclosed in a large polythene bag and closed with a tie-on label which bears the species name.

  The bags are then arranged in the normal herbarium sequence and stored on trays in the cupboards. There is, however, the possibility of damage due to crushing as more material is added to the collections.

- ***Standard mounting*** on sheets and boards may be necessary if the conditions are humid and the expense of palm-folders and boxes cannot be met. Palm-sized sheets and boards as well as standard ones will be required, and also large carpological boxes to contain fruit and bulky inflorescences. It is important that all portions removed to the ancillary collections are cross-referenced.

Fungi, lichens and myxomycetes (Slime Moulds)

See also **33. The collection and preservation of fungi and lichens**, *Principal types of larger fungi.*

Dried fungi are particularly susceptible to insect attack, so before placing them in the herbarium they must first be disinfested, preferably by deep freezing for at least 48 hours (see p. 25). Chemical fumigation is occasionally carried out (see **4. Pests and treatments**). Lichens are less susceptible to insect attack, but they should also be frozen before incorporation into the herbarium. Chemical fumigation of lichens should be avoided.

Specimens should never be pressed or mounted on sheets. In the herbarium, dried collections should, where appropriate, be placed in packets, preferably polyester (alternatively glassine) within paper packets.

Paper packets should be made of stiff cartridge paper in three sizes: e.g. 12 × 9 cm., 15 × 11.5 cm., 18 × 12.5 cm. The following designs are required:

- Envelope-type made of stiff paper (also known as moss packets), for general use. When the material is delicate and subject to crushing, e.g. small agarics, it should be placed in small, shallow cardboard boxes (7.5 × 4.5 × 1 cm.) within the packets, or the boxes can be stored separately.

- Open-topped envelopes or paper capsules (see fig. 5), for leaf parasitic fungi, e.g. rusts and smuts.

The data labels are mounted on each packet together with any determination slips and red type labels where appropriate. If the data labels are large or there are drawings of fresh fruit-bodies etc., they can be placed inside the packet and a label bearing only the minimum data added outside. Spore deposits (prints) and drawings of microstructures observed under the microscope may also be stored in the packet. This should not, however, be done for collections of delicate taxa.

Being so fragile, **myxomycetes** should be placed in shallow boxes, glued to the bottom, and all labels etc. glued to the outside.

Ideally, packets should be loose-filled in drawers or boxes; this gives better protection to the specimens and facilitates curation. Most fungi and lichens should be filed vertically like a card-index, but because of their fragility it is important that myxomycetes are kept horizontal. Alternatively the packets can be glued to standard herbarium sheets, using an approved adhesive. One sheet can accommodate up to eight small packets or four large ones, thereby saving space.

Ancillary material may be stored as follows; all material placed in separate collections must be clearly cross-referenced (see fig. 17 and p. 63):

- *Bulky specimens* which are unsuitable for storing in packets should be placed separately in carpological boxes containing a label. Extra-large specimens may be stored in polythene bags, preferably heavy-duty with a clip top, with a duplicate label inside. A portion of the specimen may also be placed in a labelled packet and mounted on a sheet.

- Material received in *spirit* should be stored separately and *never* dried out and placed in packets.

- *Permanent microscope mounts* may be stored as a separate slide collection and cross-referenced on the packet, or be kept in individual, rigid cardboard containers and placed within the packet containing the specimen.

- *Transparencies, prints and paintings* of specimens are very important in fungi as precise colour information is vital. These items should be stored separately and cross-referenced as appropriate (see **13. Collections and illustrations and photographs**).

Arrangement of the collections

If the specimens are stored on sheets, type and particularly valuable collections are placed in type covers. Sheets are placed within named species covers and the genus and species covers are taxonomically arranged in genera and families. Family, genus and species covers should be numbered, and an index to taxa placed at the beginning of each group. A geographical arrangement within the species cover is important for larger collections.

If the specimens are filed vertically or stored horizontally the individual drawers must be clearly labelled to show family and genus; dividers with indication labels may be inserted in the drawers where appropriate. It is helpful to add the family and genus numbers to each packet or box.

An example of a family arrangement is included in Ainsworth et al. (1973). However, the classification of the Fungi (incl. Lichens) is currently in a state of uncertainty and there is no single generally accepted system. Kew has adopted an expanded form of the Ainsworth general purpose classification to replace the older system of Saccardo (1882–1931).

Collections of host-specific plant parasitic fungi, such as rusts, smuts, powdery mildews and sooty moulds, may be filed according to the family of the host species. This facilitates identification but does not reflect the taxonomic relationships of the fungi.

REFERENCES

Ainsworth et al. (1973)
Hawksworth et al. (1983)
Rosentreter (1988)
Saccardo (1882–1931)
Savile (1973), chapter 2, fungi

18. DUPLICATE DISTRIBUTION

The extraction of duplicates is dealt with in **8. Processing unmounted specimens**, and **30. Collecting and preserving specimens**, (*Selection of duplicate sets*), and duplicate dispatch in **7. Centralized accessioning, recording and dispatch procedures**.

Generally speaking it is important that duplicates are distributed as soon as convenient for the following reasons:

- They become available to as many researchers as possible.
- If a disaster (e.g. fire or flood) befalls the distributing herbarium much scientific data will be saved in the recipient herbaria.
- Undistributed duplicates require safe storage space (often costly and needed for other material) and can deteriorate, especially if kept too long in tropical conditions (e.g. mould, insect damage).

Un-named duplicates

It is probably better for special and local herbaria (especially new ones) to distribute duplicates before they are named and then to correlate the received determinations with their own top set. General and national herbaria may wish to send certain taxa to specialists for critical determination. (The terms special, local general and national herbaria are defined in **2. The development, purpose and types of herbaria**). If specimens have to be sent for naming to an institution or individual where no previous agreement exists, it is courteous to write first and seek their co-operation.

Named duplicates

General and national herbaria often do not distribute duplicates until the specimens have been named. The value of a duplicate is greatly enhanced if it has been accurately named by an expert working with direct access to authentically named and well curated collections.

Distributing duplicates

It can be a problem to decide where duplicates should be sent; usually there is insufficient material to satisfy either the goodwill the donating herbarium would like to create, or the requirements of all the herbaria interested in receiving duplicates. It may be helpful to consider the following points:

- As far as possible any existing exchange agreements, often referred to as 'continuation of exchange', should be honoured. Herbaria usually have a list in descending order of institutions which are to receive the 1st, 2nd, 3rd duplicates etc. of a particular collection. In large general herbaria there will usually be separate lists for each geographical area. These lists should be open to review at regular intervals, e.g. every 5 or 10 years.

- Where a specialist has a particular interest in a family or genus it may be desirable to send a duplicate even if it means deviating from the normal distribution sequence.

- Because of better long-term storage conditions in temperate areas, it is advisable for herbaria in tropical areas, especially humid ones, to send at least one duplicate to a major general herbarium. This procedure also has the advantage that the specimen will be more accessible to scientific research, the numbers of visitors and loans being far greater at a major institution (Steenis 1957).

- Local or special herbaria require only what is relevant to their field of study.

- National herbaria are greatly enriched by well named specimens from adjacent or phytogeographically similar areas.

- When collecting in another country, it is essential to donate a complete set of duplicates to their national herbarium. In some instances this may be obligatory, but it is essential for building co-operation and goodwill.

- If a new or additional institution asks for duplicates and insufficient are available (a frequent problem for general herbaria where the main intake of specimens usually depends on duplicates received) some other service can be offered instead, e.g. prompt return of determinations for duplicates received, photographs of type specimens or reprints and photocopies.

- Some herbaria keep records of the numbers of duplicates they distribute to other institutions and expect an equal number in return. This procedure is unscientific and should not be encouraged. The scientific value of the specimens received must be considered, e.g. repeat collections from student field trips to the same area or collections of roadside weeds will have very little value compared with one rarity with good data from a remote and unfrequented area.

Recording duplicate distribution

This information can be important if the specimen is to be cited in a flora or monograph, especially for tracing isotypes (if a specimen is described as a new taxon).

- Collectors sometimes personally keep distribution lists in their collecting books or with their determination lists.

- Some institutions print the standard herbarium abbreviations of recipient herbaria at the foot of the data labels and then encircle the herbaria as

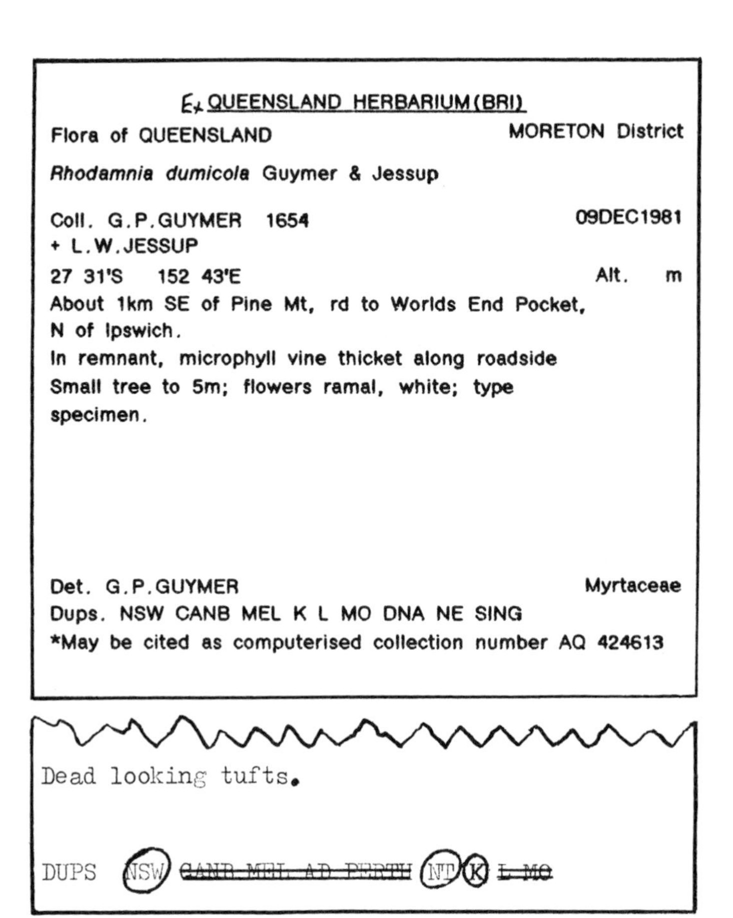

Ex QUEENSLAND HERBARIUM (BRI)

Flora of QUEENSLAND MORETON District

Rhodamnia dumicola Guymer & Jessup

Coll. G.P.GUYMER 1654 09DEC1981
+ L.W.JESSUP

27 31'S 152 43'E Alt. m

About 1km SE of Pine Mt, rd to Worlds End Pocket,
N of Ipswich.
In remnant, microphyll vine thicket along roadside
Small tree to 5m; flowers ramal, white; type
specimen.

Det. G.P.GUYMER Myrtaceae
Dups. NSW CANB MEL K L MO DNA NE SING
*May be cited as computerised collection number AQ 424613

Dead looking tufts.

DUPS NSW ~~CANB MEL AD PERTH~~ NT K ~~L MO~~

FIG. 27 — Data labels including records of duplicate distribution.

applicable (see fig. 27). This also prevents any subsequent duplicates extracted by the recipient herbaria being distributed to an institution that has already received one.

— Some institutions keep ledgers or proforma sheets for recording duplicates distributed from their herbaria (see example in Womersley 1981: 59).

REFERENCES

Fosberg & Sachet: 96–97 & 103 (1965)
Steenis, van: LXV (1950)
— : 133–135 (1957)
Womersley: 59 (1981)

19. LOANS TO OTHER INSTITUTIONS

The handling of loan material is an important aspect of a technician's duties, and the reputation of the institution depends in part on the efficiency of its loan service. Loans of specimens to and from herbaria are frequently arranged because:

- It is essential for a taxonomist preparing a revision to examine the type specimens, and it is highly desirable that as much material as possible is studied. It is not always possible to visit all the herbaria concerned.

- Direct comparison can be made between specimens present in the borrowing institution and specimens borrowed from other institutions.

- It is in the interest of the lending institution to have its specimens annotated by specialists.

Requesting loans

Loans are usually officially made between institutions and *not* individuals. Correspondence should be between the Director or Curator of the borrowing institution, on behalf of the botanist needing the material, and the Director or Curator of the lending institution. Since loans are costly both in terms of postage and packing and in the amount of staff time needed to process them, requests should be kept to a reasonable size.

It is important that requests include precise information about the specimens needed. Species names and synonyms, country or region, collectors' names and numbers and type status should all be included as far as possible. The request should be listed in a logical order so that items are not overlooked by the person extracting the specimens. If the letter is inadequate the sender should be asked for the necessary information.

If the loan material has been personally selected by a botanist while visiting the lending institution, some kind of formal request may still be necessary. Larger herbaria often provide application forms for this purpose. After completion the form should be approved by the head of the visitor's institution before the loan is dispatched, since the borrowing institution has to accept responsibility for the care of the specimens while they are on loan.

Unsolicited loans

If an institution wishes to send particular specimens to a specialist, permission from the receiving botanist or institution must be sought in advance.

Policy concerning loans

Herbaria should, as far as possible, make their collections available for scientific study by complying with reasonable loan requests, but in so doing some degree of risk is inevitably involved while the collections are out of the lending institution's custody.

Many herbaria have a stated policy and a printed list of regulations concerning the loan of specimens. Such documents are generally compiled with the safety and preservation of the collections in mind. The following points should be noted when considering policy and regulations:

- Clearly state any restrictions concerning the removal of parts of the specimen for study.
- If the loan request is from a politically unstable country, or where the postal services are unreliable, the increased risk of loss must be considered. Good photographs are often adequate and can be sent instead of a loan.
- Pack the specimens well for transit and request their return in the same or similar materials.
- It is important that the borrowing herbarium does its utmost to ensure the safety of specimens. All loans should be stored safe from insect attack and handled with care at all times.

Any damage that may have occurred in transit should be noted and the sending institution informed immediately.

If the material, or one parcel from a consignment, does not arrive at the expected time, notify the sending institution so that they can make enquiries at the post office.

Selecting material to be sent on loan

When a loan request has been approved by the Director or Curator the material must be removed from the collections, bearing in mind any guidelines or restrictions that may be institutional policy.

If a requested specimen cannot be found, a botanist should be consulted; if any such specimens still cannot be traced, this fact should be communicated to the botanist requesting the loan.

If only a proportion of the material available is to be sent on loan:

- The full range of variation for each taxon including flowers (both ♂ & ♀ if the plants are unisexual) and fruits must be selected.

— Widely distributed collections already available to the borrower need not be included.

If data such as country or accepted specific name are not shown on the herbarium sheet add them in pencil. This will make it easier to refile the sheets on their return.

A hanging label with details of the loan should be placed in the herbarium to indicate that specimens have been removed. This can save a lot of time by preventing useless searching for absent material, and it acts as a warning not to fill the gaps with new acquisitions or by spacing out surrounding material.

Necessary repairs and cleaning should be attended to before sending on loan. (See **10. Conservation of sheets**).

The sheets should be protected by placing them in double flimsies or newspaper. Packing such as foam or wadding should be used to help support and protect bulky or thick specimens before bundling.

Spirit or carpological material should be included as appropriate. Spirit material is best dispatched in special leak-proof plastic bottles (not glass) (see **13. Ancillary collections**, *Loans*).

Procedure for recording and dispatching loans

See **7. Centralized accessioning, recording and dispatch procedures**.

Material returned from loan

Any sheets damaged in transit must be repaired before returning specimens to the collections (see **16. Incorporation of mounted specimens**).

REFERENCES

Franks: 18 (1965)
Fosberg & Sachet: 95 & 101–102 (1965)
Kobuski et al. (1965)
Radford et al.: 764–765 (1974)
Savile: 54–57 (1973)

20. VISITORS

It is both customary and polite for visitors to write formally requesting permission to work in another herbarium; the reply letter granting permission can serve as identification and be shown on arrival. Notice of a proposed visit can allow staff to make any special arrangements on the visitor's behalf or to indicate any problems (e.g. local holidays during which the herbarium will be closed, or if the material required is out on loan).

If visitors wish to bring their own collections into the herbarium the possibility of infection by insects must be considered. Either fumigate or deep freeze, but if the visitor's time is short and only a few specimens are involved, microwave or carefully inspect material before it is allowed into the herbarium. A large notice by the entrance warning visitors to declare any herbarium specimens can save embarrassment. If possible tell them by letter before they arrive. Similarly it is important to check computer diskettes for viruses etc. (see **29. Introduction to computers**, *Data and system security*).

Side 1

REGULATIONS FOR VISITORS TO THE HERBARIUM AND LIBRARY

1. . . . .
2. . . . .
3. . . . .
4. . . . .

< fold

5. . . . .
6. . . . .
7. . . . .
8. . . . .

< detach

DECLARATION

I have read and understand the Regulations for Visitors to the Herbarium and Library and I undertake to observe them.

Name and Title: ....

(BLOCK LETTERS) ....

Institution or Permanent address: ....

(BLOCK LETTERS) ....

....

Date .... Signature ....

Approximate duration of visit ....

Subject to be studied ....

Side 2

[NAME OF INSTITUTION]
HERBARIUM & LIBRARY
VISITOR'S CARD

Valid until ....

Holder ....

Holder's Signature ....

Authorized by .... Date ....
Keeper/Chief Librarian

Please have this card available when entering the building.

here >

9. . . . .

10. . . . .

here >

Address during visit: ....
(BLOCK LETTERS)
....
....

Telephone No.: ....

Botanist reponsible ....

FIG. 28 — Suggested design for a visitor's card.

A *Visitor's Card* may be found useful, especially in larger institutions. The card can have two functions (see fig. 28):

1. The visitor's name, institution or permanent address, address during visit and subject of study are entered by the visitor on a detachable portion of the card which is filed for reference. This portion also bears the signed declaration by the visitor that the regulations will be observed.

2. The main part of the card bears the visitor's name and is printed with the herbarium regulations. It is signed by the head of the herbarium and carries an expiry date; it can thus be used as a pass.

Each day visitors should be asked to sign a *Visitor's Book* as they arrive. This forms an interesting archival record and can provide useful statistical data such as the identification of the busiest times of the year; it can also be used as a security check.

Most herbaria give suitable visitors free access to the collections under the supervision of a staff member. Some herbaria prefer to restrict visitors to a separate work area and arrange for the technical staff to bring the required material to them. Whether or not a visitor should be given free access to the collections will depend on how experienced he or she is in herbarium work, and on the judgement of the staff member responsible. If necessary, a visitor can be requested *not* to return material to the cabinets.

If free access is given, visitors will need to know how to find their way around the herbarium and locate required taxa. A member of staff should be available to conduct visitors around the building and explain the organization of the collections. In addition to a personal briefing it is worth considering the compilation of a printed *guide to the herbarium* for the use of visitors.

If the arrangement is systematic, either indicate where copies of the index to families can be found or include a family index in the guide. Also indicate where generic indexes can be found and, if necessary, species indexes.

Not all visitors to a herbarium are botanists; many are unfamiliar with herbarium procedures and do not appreciate the importance of handling specimens with the utmost care. If no visitor's cards of the type described above are available, it is a good idea to have a printed notice of the rules and regulations of the herbarium to show visitors before they are admitted. Even so, some training in specimen handling and use of the herbarium may be required. Where appropriate the visitors' attention should be drawn to the main points listed under **15. Handling herbarium specimens**; a specially prepared hand-out could be made available.

21. REMOVAL OF SAMPLES FROM HERBARIUM MATERIAL

With the growing importance of anatomy, cytology, palynology and phytochemistry in relation to taxonomy, herbaria receive increasing numbers of requests for portions of material for such research. Where appropriate, requests for samples should first be made to institutions in the applicant's own country or countries where the plants concerned grow. If only very little material of a species is present in a herbarium it may not be possible to spare portions of it. But if adequate material is present, especially in a local or national herbarium, then in the interests of science portions should be supplied if the request is considered to be worth supporting. However, the larger general and national herbaria have a particular responsibility to the international botanical community to preserve their collections for taxonomic studies and not to allow material of the rare species to be depleted. Visitors to herbaria are usually expected not to remove portions of specimens without permission; any rules concerning this and the procedure to be followed should be explained to visitors as soon as they arrive so that potentially embarrassing situations can be avoided (see **20. Visitors**).

The following general points should be borne in mind:

- It is not advisable to remove samples from types or from specimens of historical importance.

- Samples, when taken, should be from specimens in good condition and bearing adequate data.

- Suitable loose material may be present in the paper capsules, avoiding the need to remove portions from the mounted specimens.

- Place the material inside a small packet. Cellophane or Glassine is suitable, especially for pollen, as loose grains may stick to rough non-glossy paper packets or to polythene by static electricity. On the packet, or on an outer, larger, stronger packet or envelope, write the name of the plant, the collector's name and number with date, and the locality.

- Attach a small label to the sheet indicating the kind of sample removed e.g. leaf, stem, buds, stamens etc., and the name and location of the recipient with date (see fig. 11, p. 49). This information is useful if a further request is received for a similar sample from the same species, in which case the correspondent can be told where a sample has already been sent so that he or she can possibly borrow a slide or request relevant data from the original recipient.

- When sending samples, it may be desirable to ask for duplicate slides and/or light or SEM (scanning electron microscope) micrographs and any other data in exchange. This will depend on how useful these items would be to the herbarium supplying the samples. Similarly, copies of any publications citing material supplied can be requested.

— It is useful to keep a record of what has been sent and to whom for future reference; this information can be incorporated with records of herbarium loans.

— Samples must never be removed from herbarium sheets which are on loan without first obtaining permission from the lending institution.

SELECTING POLLEN SAMPLES

First of all, check if the species concerned is dioecious (i.e. the sexes are on separate plants), in which case a male specimen is needed. In monoecious species (both sexes on the same plant), if the flowers are unisexual a male flower must be found. Pollen is best obtained from mature buds; open flowers may have dehisced anthers or may have been contaminated by pollen from another plant.

Choose a specimen with plenty of buds and flowers; if the specimen has very few flowers or buds these are best not removed in case the overall scientific value of the herbarium specimen is diminished.

To take a pollen sample: remove a bud or, if no buds are available, a few indehisced anthers (check with a hand lens) with fine forceps.

If bracts are concealing possible buds on a specimen, remove a bract, check that buds are present, remove some and place the bract in a paper capsule. If the bracts are in a complex arrangement, e.g. a catkin, make sure that the value of the herbarium specimen is not reduced by damaging the only catkin. If there is plenty of material, break off a portion of the catkin, remove the buds required, and place the remaining portions in a paper capsule.

If slides are requested in exchange, state that they should be prepared by the Erdtman Acetolysis method. (see Le Thomas 1989).

For further details of pollen preparation see Faegri & Iversen 1989.

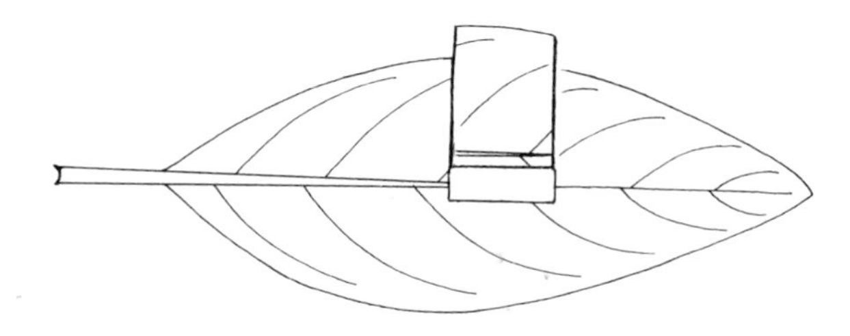

FIG. 29 — Large leaf showing position of cuts for removal of sample for anatomical study.

SELECTING ANATOMICAL SAMPLES

Leaves: if the leaves are small, remove one; but if the leaves are large, cut out a strip c. 1.5 cm. wide from the margin to the midrib inclusive, at the middle of the leaf (see fig. 29). It is preferable but not essential to include the tip and base of monocotyledonous leaves. Do not remove the youngest, smallest or largest leaf, the full range of variation should be left on the sheet; also if only one leaf shows the undersurface do not remove it. For studies of the **leaf surface**, e.g. hairs and cuticle: portions that are free from glue should be chosen. The alcohol used in collection by the Schweinfurth method (see p. 223) or carriers in liquid insecticides could dissolve the waxes of the cuticle and spoil the structure; specimens thought to have been subjected to such treatments should be avoided.

Stems: for wood anatomy, choose the oldest twigs available, preferably at least 1.5 cm. in thickness. Using a strong, sharp scalpel or single-edged razor blade, remove as carefully as possible a 2–3 cm. length (1 cm. would be acceptable if this is not possible). If the stem is thicker than 3 cm. a portion would suffice. For non-woody stems, the anatomist may require nodal or internodal material, according to the type of investigation.

SELECTING SAMPLES FOR PHYTOCHEMISTRY

Select material which has been air-dried rather than preserved in solvents since the latter procedure may have caused leaching of chemicals of interest: flavonoids and other phenolics are known to be susceptible (see Cooper-Driver & Balick 1979 & Coradin & Ginnasi 1980). Ideally any specimens collected by the Schweinfurth method or otherwise treated with preservatives should be labelled as such. In practice this has seldom happened, but dark brittle specimens can indicate alcohol collection and these should be therefore avoided.

The level of a particular chemical which can be detected in herbarium specimens depends on:

- The level present when it was collected.
- The preservation method used.
- Whether there was any change during drying.
- Whether there has been further deterioration in storage.

Despite these variables, successful chemical analyses have been made on material 50 years old and older. Some assessment of the survival of a particular compound may be made by comparing the levels in a freshly picked plant with those in the dried specimen.

The quantity of dried material adequate for analysis will depend on both the level of compound and the sensitivity of the assay system being used. When examining plant material for known classes of compound using the most modern analytical techniques only a few mg. (c.4 $cm.^2$ of leaf material) may be necessary.

If possible inform the chemist which chemicals have been used in the mounting and curatorial procedures (e.g. glue, insecticides, fungicides) so that these may be taken into account when designing the assay or analytical systems.

SELECTING CYTOLOGICAL SAMPLES

Cytologists carrying out karyological investigations prefer to work from fresh samples, but provided the material is still alive, pollen can be germinated in weak sucrose and mitosis studied in the pollen tubes. See above for selection of pollen samples.

Samples of leaf material (1–5 gms.) can be used for molecular studies (DNA extraction). Young leaves are preferred; any suspected of being collected by the Schweinfurth method and material poisoned by the application of mercuric chloride must be avoided.

SELECTING SEED SAMPLES

Seeds may be required for anatomical or phytochemical investigation. The anatomist may wish to study either the general morphology of the seed or the fine detail of the seed-coat (or testa), in either case only 1–few seeds will be needed. The phytochemist will require sufficient seeds to weigh a few mg.

Select only fully mature, well formed seeds. If a herbarium sheet bears only one large intact fruit, do not open it just to satisfy a request for a seed sample. As with pollen samples, microscope slides and SEM prints should be requested in return.

REFERENCES

Cooper-Driver & Balick (1978)
Coradin & Ginnasi (1980)
Faegri & Iversen (1989)
Le Thomas (1989)

22. ESSENTIAL HERBARIUM LITERATURE

The literature needed in a herbarium will depend on the geographical coverage of the collections and the work being carried out. In this section only the main works useful to herbarium techniques are listed.

INDEX HERBARIORUM

Holmgren, P.K. et al. (1990). **Index Herbariorum.** 8th ed. New York Bot. Gdn., New York (Regnum Vegetabile vol. 120).

Information on herbaria arranged alphabetically by countries then by cities or towns, with full addresses, lists of staff members, specializations in research, principal collections, loan policy, publications etc. Important for arranging loans and for communication. Includes herbaria codes (e.g. K = Kew etc.) which are now in almost universal use in taxonomic literature, and found on many herbarium labels.

DICTIONARIES OF PLANT NAMES

- **Index Kewensis**. Original work plus 19 supplements. Includes all binomial and generic names of flowering plants from Linnaeus (1753) onwards and all names at rank of family and below from 1971. Gives place of first publication and area of origin and, in the case of new combinations, the basionyms. Supplement 19 includes entries for 1990. Oxford University Press plan to publish a computer-readable version of the Index Kewensis on a CD-ROM (compact disk) in 1993. It will include the entire work as well as records for 1991.

  Owing to changing editorial policy over the years, Index Kewensis must be used with caution. Names listed are not necessarily valid and the author citations may be incomplete or incorrect. The indications of synonymy which were given until Supplement III should be disregarded: they are taxonomic *opinions* and not bibliographical *facts*, which are the main concern of the work. Occasionally, the spelling of a name, although faithfully copied from the original publication, may not be the correct spelling according to the current rules of nomenclature. Index Kewensis must therefore not be used as an authority in any sense, but rather as a *guide* towards tracing where a name was first published. See Meikle (1971) and the Introduction to the CD-ROM version of Index Kewensis.

- Farr, E.R. et al. (1979). **Index nominum genericorum**. Utrecht, plus Supplement 1 (1986). Lists all generic names of plant and fungal groups with place of publication and type.

- Mabberley, D.J. (1989). **The plant book**. Cambridge University Press, Cambridge. Comprehensive portable dictionary of flowering plants, including conifers, and ferns. First printed in 1987, the second edition has many corrections. Reissued 1992.

- Willis, J.C. (1973). **A dictionary of the flowering plants and ferns.** 8th edition by H.K. Airy Shaw. Cambridge University Press, Cambridge. Gives names, authorities, number of species and geographical range of vascular plant genera, and descriptions and internal classification for family names.

- **Index Filicum** (1906 to date). Original work plus 5 supplements (supplement 6 is due in 1993). Includes all binomial names of ferns from Linnaeus onwards and of all pteridophytes from 1961. Gives full basionym references and will list all infraspecific ranks in future supplements.

- **Index Muscorum** (1959 to date). Gives all names of mosses with basionyms, synonyms and places of publication. Updated in 'Taxon'.

- Brummitt, R.K. (1992). **Vascular Plant Familes and Genera**, Royal Botanic Gardens, Kew. Lists accepted generic names and selected synonyms in an alphabetical list indicating their families. Lists accepted families alphabetically with enumeration of their accepted genera. Also indicates placement of flowering plant families in eight systems of classification (Bentham & Hooker, Dalla Torre & Harms, Melchior, Thorne, Dahlgren, Young, Takhtajan and Cronquist), whose sequences are all listed.

- Bonner, C.E.B. (1962–78). **Index Hepaticarum**, Parts 1–9, Cramer. Includes names of hepatics at all ranks with places of publication, basionyms, details of typification and geographical distribution.

- **Index of fungi**. (1940 to date). Supplements twice a year. CAB International Mycological Institute, Kew. Preceded by Petrak: Index of Fungi, 1920–1939; Saccardo: Sylloge, 1882–1931.

- Other, regional indexes: for example: **AETFAT Index** (1954 to date); covers tropical African names of flowering plants of all ranks. **Gray Herbarium Index** (1894 to date on cards until 1984 then on microfiche); covers all North and South American names of flowering plants at all ranks from 1885 onwards.

MAPS, ATLASES AND GAZETTEERS (see **26. Collectors, itineraries, maps and gazetteers**)

TAXONOMIC BIBLIOGRAPHIES

- **Kew Record of Taxonomic Literature** (1971 to date). HMSO, London. Since 1987 this has been issued quarterly. Covers the world literature on vascular plants.

– **Bibliography of Systematic Mycology** (1943 to date). CAB International Mycological Institute, Kew. Published twice a year.

– See also works listed above.

FLORAS, IDENTIFICATION MANUALS AND GLOSSARIES

– **Floras and manuals**
The choice of floras required at local, national and regional levels will depend on the needs of the individual herbarium. The coverage is too large for inclusion here, but is dealt with by Frodin, D.G. (1984), **Guide to standard floras of the world**, Cambridge University Press, Cambridge.

– **Glossaries**
Jackson, B.D. (1928). **A glossary of botanic terms**. Duckworth, London (last reprinted 1971).

Stearn, W.T. (1983). **Botanical Latin**. David & Charles, Newton Abbot (England). 3rd ed. Has useful line drawings illustrating botanical terms.

AUTHORS OF PLANT NAMES

– Stafleu, F.A. & Cowan, R.S. (1976–1988). **Taxonomic Literature** (ed. 2) 7 vols. Bohn, Scheltema & Holkema, Utrecht. This is a comprehensive guide to taxonomic literature from 1753 to 1939 arranged by authors. Information is given on authors, their publications with dates, and the locations of their herbaria and types.

– Brummitt, R.K. & Powell, C.E. (1992). **Authors of Plant Names**. Royal Botanic Gardens, Kew. Lists authors of names of all plant groups with abbreviations and dates of the authors. Supersedes Meikle, R.D. (1984), Draft Index of Author Abbreviations, HMSO which only included authors of flowering plants.

ABBREVIATIONS

Certain abbreviations (or terms) frequently found in literature citations may be unfamiliar. The most commonly encountered are:

f. (or **fig.**) (*figura*) = figure or illustration.
ib. (or **ibid.**) (*ibidem*) = the same or in the same place.
ic. (*icon*) = illustration.
id. (*idem*) = the same.
in litt. (*in litteris*) = in correspondence.
in sched. (*in schedule*) = on a herbarium label.
l.c. (or **loc. cit.**) = (*loco citato*) = at the place cited.

MS (or **MSS**) (*manuscriptum* (or *a*)) = manuscript(s).
op. cit. (*opere citato*) = in the work cited.
t. (or **tab**.) *tabula* = plate.
tom. cit. (*tomus citato*) = in the volume cited.
TS = typescript.

Loc. cit. implies that the reference is exactly the same (i.e. page number) as the one cited previously; *op. cit.* implies that it is in a work of the same title but on a different page and *tom. cit.* that it is on a different page of the same volume. *Ib.* and *Id.* are less frequently used and can mean the same as *loc. cit., op. cit.* or *tom. cit. In. litt., in sched., MS* and *TS* all indicate that the information was not formally published.

ROMAN NUMERALS

Many older periodicals have the volume numbers indicated as Roman numerals; often plates and occasionally the text (or part of it) can be similarly treated. The following guide is included to help those unfamiliar with this system.

Basic units

I = 1	X = 10	C = 100	M = 1000
V = 5	L = 50	D = 500	

Most numbers are formed by addition e.g.:

VI	5 + 1 = 6	XI	10 + 1 = 11
VII	5 + 1 + 1 = 7	XII	10 + 2 = 12
VIII	5 + 1 + 1 + 1 = 8	XIII	10 + 3 = 13
		XIV	10 + 4 = 14
LX	50 + 10 = 60	CIX	100 + 9 = 109
LXX	50 + 20 = 70	CCXX	200 + 20 = 220
LXXX	50 + 30 = 80	MDIII	1000 + 500 + 3 = 1503

Some numbers i.e. 4 and 9 or numbers including 4 and 9, are indicated by subtraction e.g.:

IV	= 5 – 1 = 4
IX	= 10 – 1 = 9
XL	= 50 – 10 = 40
XC	= 100 – 10 = 90
CD	= 500 – 100 = 400
CM	= 1000 – 100 = 900

Examples of numbers

I	1		XI	11	XXX	30
II	2		XII	12	XL	40
III	3		XIII	13	L	50
IV	4	(rarely IIII)	XIV	14	LX	60
V	5		XV	15	LXX	70
VI	6		XVI	16	LXXX	80
VII	7		XVII	17	XC	90
VIII	8		XVIII	18	C	100
IX	9		XIX	19	D	500
X	10		XX	20	M	1000

Additional examples and use of the 'backward C' are given in Stearn (1983) in the chapter headed 'Numerals and Measurements'.

Dates

Some older books show the year of publication in Roman numerals. This can appear quite bewildering; however, remember the years covered will generally fall between 1750 and 1900.

MDCC	1000 + 500 + 100 + 100 = 1700	
MDCCL	1000 + 500 + 100 + 100 + 50 = 1750	
MDCCC	1000 + 500 + 100 + 100 + 100 = 1800	e.g. MDCCCXV = 1815
MCM	1000 + (1000–100) = 1900	e.g. MCMLXXXVIII = 1988

REFERENCES

Cervera (1986)
Meikle (1971)
Stearn (1983)

23. REARRANGEMENT OF HERBARIUM COLLECTIONS ACCORDING TO NEW PUBLICATIONS

Since the herbarium is an important source of information about plants, it is essential that the names in use in the herbarium are kept up-to-date.

It is therefore necessary from time to time to revise the collections according to a new revision or floristic publication. Where appropriate, the advice of a senior botanist should be obtained before attempting such a task. The species may need rearranging in the order set out in the publication and an index including synonyms compiled. Even if the herbarium has an alphabetical arrangement it will still be necessary to deal with taxa placed in synonymy and note any name changes or newly described taxa.

I [**14. Keetia tenuiflora** (*Hiern*) *Bridson* comb. nov. a

II {
Canthium tenuiflorum Hiern, Cat. Afr. Pl. Welw. 1(2): 477 (1898). Type: Angola, Pungo Andongo, near Quilanga and Quibanga *Welwitsch* 3143 (syntype LISU not seen; isosyntypes BM, K) & Barrancos 3144 (syntype LISU not seen; isosyntypes BM, K).
Plectronia tenuiflora (Hiern) K. Schum. in Just's Jahresb. 1898. 1: 393 (1900).
P. angustiflora De Wild. in Bull. Jard. Bot. Brux. 5: 30 (1915); **synon. nov.** b Type: Zaire, Equateur Province, Dundusana, *Mortehan* 911 (holotype BR).
P. rutshuruensis De Wild., Pl. Beq. 3: 197 (1925); **synon. nov.** Type: Zaire, b Kivu Province, Rutshuru, *Bequaert* 6289 (holotype BR).
Canthium brownii Bullock in Bull. Misc. Inf. Kew, 1932: 370 (1932). Type: Uganda, Entebbe, *Brown* 233 (holotype K).
[*Canthium zanzibaricum* sensu Bullock in op. cit. 1932: 373 (1932) pro parte, c d f quoad *C. tenuiflora* in syn.; sensu Hepper in Fl. W. Trop. Afr. ed. 2, 2: 184 c e (1963) pro parte, quoad *Deighton* 3058 & *Small* 471, non Klotzsch]. d e f

DISTRIBUTION. Uganda (U2 & 4); Sierra Leone, Nigeria, Cental African Republic, Gabon, Cameroun, Zaire, Angola.

I [**15. Keetia koritschoneri** *Bridson* sp. nov. affinis *K. zanzibaricae* (Klotzsch) g Bridson sed foliis typice majoribus, paribus nervorum lateralium principalibus paucioribus usque 7, domatiis pubescentibus, differt. Typus: Tanzania, *Koritschoner* 1412 (holotypus EA; isotypus K).

III {
Frutex scandens vel ?arbor 7 m alta; ramuli juniores glabri. *Laminae* oblongae-ellipticae, 10–16 × 5·8–7·7 cm, apice breve acuminatae, basi ob-
. . . .
Semina endospermio granulato resinoso continentia. (Fig. 4 K–P).

IV {
TANZANIA. Lushoto District: Makuyuni District, fr. June 1935, *Koritschoner* 669 (EA), fl. (mostly fallen)—1935, 1327 (EA, K) & fl. mostly fallen—1935, 1412 (holotype EA; isotype K); Mombo Forest Reserve, 400 m, young fr. 6 Jan. 1965, *Muze* 18 (EA).

FIG. 30 — Example of a revision taken from Kew Bulletin. **I** correct species name and number; **II** synonymy; **III** description; **IV** specimen citations. Note the use of: **a** *comb. nov.*; **b** *synon. nov.*; **c** *sensu*; **d** *pro parte*; **e** *quoad*; **f** [ ]. and **g** *sp. nov.*

Specimens which have been on loan to the author should bear his or her determination labels, and 'seen for ...' labels in the case of a Flora account (see fig. 10, p. 49). Specimens not annotated by the author may appear in the lists of specimens cited (see below, *Citations*).

USING A REVISIONARY OR MONOGRAPHIC WORK

Revisions vary greatly in the amount of detail and format (arrangement of the information) used. They may contain much discussion and ancillary information (e.g. palynology, anatomy etc.) that is not strictly relevant to the task of curation; however, the central part of the work should comprise a sequenced list of all the taxa the author recognizes, setting out the basic information as follows (see fig. 30):

- species number (unless arrangement is alphabetical) and species name (see fig. 30, I).
- list of names included in the synonymy (see fig. 30, II).
- description of the species (not always present) (see fig. 30, III).
- citation of specimens (often with notes on distribution and ecology (see fig. 30, IV).

The correct species name and any names placed in synonymy will be followed by references to other works in which they have been used. If any of the bibliographical abbreviations are unfamiliar see **22. Essential herbarium literature**.

Note: the examples given below are for species rank but they could also be applied to other ranks.

The correct species name

First take note of the species name and number and trace any material under this name in the herbarium. Take special note of any names followed by the terms **sp. nov.**, **nom. nov.**, **comb. nov**. or **stat. nov**. (the latter especially at infraspecific level) (see *Glossary*). In order to confirm that the specimens are correctly named check for cited material (see below, *Citations*). If no cited specimens are present a botanist can then compare the material with the descriptions or use the key.

Names in synonymy

Usually the synonyms are listed in strict chronological order beneath the correct species name. Specimens of all the taxa placed in synonymy should be found and put under the correct epithet. The list of synonyms may include names that are not strictly synonyms and it is important that the following points are understood before specimens are transferred and synonyms included in indexes:

Synonym. An incorrect name for a taxon; usually a legitimate name that was validly published but later than the correct name.

Misidentification. A name which has been applied to the wrong taxon, e.g.:

— *Rosa alba* sensu Green, non White – i.e. Green wrongly applied the name '*Rosa alba* White' to material of a different species (the species to which the present author is referring).

— *Rosa alba* sensu auct., non White – i.e. many authors have used the epithet for specimens belonging to the present species but not to White's *Rosa alba.*

— *Rosa alba* sensu Pink, pro parte, non White, quoad *Ford* 1, *Jones* 2 and *Lee* 3 – i.e. Pink used this name for a mixture of species partly not belonging to White's *Rosa alba.* Of the specimens Pink cited, only *Ford* 1, *Jones* 2 and *Lee* 3 should be placed in the present species.

Unavailable names

a. *Invalid name.* A name which has not been validly published according to the International Code of Botanical Nomenclature (ICBN) and therefore cannot be used as the correct name for a taxon, e.g. **nom. nud.** and **ms** (see below, *Glossary*).

b. *Illegitimate name* (see '**nom. illegit.**' in *glossary*).
e.g. **homonym**, such as *Rosa rubra* Green (1960) non Blue (1955) – i.e. Blue validly published the name *Rosa rubra* as a new species in 1955. Later, in 1960, Green used the same name when publishing another new species. Since the same name cannot be used for more than one species (or more than one taxon at any other level) *Rosa rubra* Green must be rejected, being a *later homonym* (= same name). If no other validly published name for Green's plant is available a new name will be required (see '**nom. nov.**' in *glossary*).

Type specimens

These are usually cited after both the correct name and synonyms (see fig. 30, I & II). Any such specimens should be noted and type labels (see fig. 12) applied as appropriate. Care should be taken to select the correct kind of type label; if in doubt seek advice. Simple definitions of the more important kinds of type used at species and levels below species are given below.

holotype – the specimen (or sole element) of a collection designated (or indicated) by the author at the time of publication as the type specimen of the name of a species (or subspecies or variety).

isotype – any duplicate of the holotype, i.e. part of the same collection.

syntype – any one of two or more collections cited by the author, at the time of publication, when no type was designated; or any one of two or more specimens designated as types.

lectotype – a specimen chosen by a subsequent author from amongst syntypes to serve as the definitive type.

neotype – a specimen chosen by a subsequent author to act as a type when the original material has been lost.

All type specimens should be protected by placing inside a specially designed type-cover, or separate species-cover if a type cover is unavailable (see fig. 25).

Citations

The description of the species (if present) will usually be followed by citations of specimens. There may be a complete list of all the specimens seen by the author or just a representative sample. Specimen citations are usually arranged in geographical order with collections from the same region alphabetically arranged by collector. The standard abbreviations for herbaria holding the specimen (or duplicates) are included in brackets after the collector's name and number (see fig. 30, III). Sometimes the author will indicate whether or not the specimen has been seen (see below, *Glossary*, '**!**' and '**n.v.**'); if there is no apparent indication there may be a statement in the introductory remarks such as 'all cited specimens have been seen unless otherwise stated'.

Sometimes there is a combined list of all the specimens of all the species seen by the author in alphabetical order of collectors, together with an indication of the identifications, usually referred to as '*list of exsiccatae*' or sometimes just *exsiccatae* (see below, *Glossary*). An identification list of this kind may be included at the end of the work or it may be issued as a separate publication (sometimes only available on request). Any cited specimens in the collections should be noted and named as indicated. The determination is written on a determinavit label with the indication 'cited in ...' or 'ex num.' (see below, *Glossary*).

Indexes

An index including both correct names and synonyms should be compiled and placed in the herbarium. It is important that the synonyms are clearly distinguished from the correct names, e.g.:

alba Smith 1
pallens Brown . . . . = 1
or pallens Brown . . . . = alba

It is useful, but not essential, to include authors. However, if homonyms or mis-used names are to be indexed these must be shown with the author citations to prevent confusion with true synonyms or accepted names, e.g.:

rubra Brown . . . . .2 *or*: rosea Brown 5
rubra Smith, non Brown . . . = 4 rosea sensu Smith, non Brown . . . = 6

If the index in the published work is clearly set out with species numbers it can save time to photocopy it instead of compiling a new one. The index must be clearly titled with the generic name, geographical area and the bibliographical reference of the arrangement, e.g.:

'Index to the species of *Psydrax* in Africa, arranged according to Bridson in Kew Bull. 40: 687–725 (1985)'.

It is useful to include a photocopy of the key with the index.

USING A FLORISTIC WORK

This is essentially similar to curating from a revisionary work, but it can be an easier task. Many floras do not formally publish new taxa or name changes. If necessary, these will have been dealt with in preliminary papers published in a

I **1. S. africana** *Sond.* in Linnaea 23: 106 (1850); Pax in E.P. IV. 147(5): 155 (1912); Prain in F.T.A. 6(1): 1006 (1913); V.E. 3(2): 139 (1921); T.T.C.L.: 226 (1949); K.T.S.: 220 (1961). Types: South Africa, Cape of Good Hope, Sterkstroom [Sterkstrom], *Zeyher* 1528 (S, syn.!, ? GRA, K!, P!, SAM!, etc. isosyn.) & Natal, Durban [Port Natal], *Gueinzius* 173 (? S, syn., G, MEL, isosyn.!, K, photo. of isosyn.!) & 174 (S, syn.!) a a a

III A dioecious or monoecious several-stemmed, much-branched deciduous shrub or tree with a rounded crown (2.5–)4–9(–14) m. high, often with drooping stems and
. . . .
. . . .
brown or light brown, streaked with reddish brown or dark brown.

IV KENYA. Kwale District: Mteza, Aug. 1937, *Dale* in *F.D.* 3769!; Teita District: Galana R., Lugard Falls, 1 Jan. 1967, *Greenway & Kanuri* 12923! & near Sobo Rocks, 9 Jan. 1967, *Greenway & Kanuri* 13012! a a a
TANZANIA. Lushoto District: Mkomazi, 6 Sept. 1968, *Shabani* 202!; Morogoro District: 5 km. S. of Mziha R., 20 Nov. 1955, *Milne-Redhead & Taylor* 7096!; Lindi District: without precise locality, or date, *Gillman* 1520! a a a

DISTR. **K** 7; **T** 2, 3, 6, 8; Mozambique, Zimbabwe, Botswana, Angola, Namibia, Swaziland and South Africa (Transvaal, Natal, Cape)

HAB. Locally common in deciduous woodland, bushland and wooded grassland; 15–900 m.

II SYN. *Stillingia africana* (Sond.) Baill., Etud. Gén. Euph.: 522 (1858)
Excoecaria africana (Sond.) Muell. Arg. in Linnaea 32: 123 (1863) & in DC., Prodr. 15(2): 1215 (1866)
Maprounea africana Muell. Arg. in DC., Prodr. 15(2): 1191 (1866), pro max. parte
Excoecaria synandra Pax in E.J. 43: 223 (1909). Type: Tanzania, Morogoro, *Holtz* 1735 (B, holo. †) b
Excoecariopsis synandra (Pax) Pax in E.J. 45: 239 (1910)
Spirostachys synandra (Pax) Pax in E.P. IV. 147(5): 155 (1912); V.E. 3(2): 138, 139 (1921)

NOTE. The reference to a gathering from **K** 1 in K.T.S. is erroneous; this belongs to the next species. *S. africana* has so far not been found N. of 3° S.

FIG. 31 — Example from Flora of Tropical East Africa. **I** correct species name and number with references and type information; **II** synonymy; **III** description; **IV** specimen citations. Note use of **a** '!' and **b** '†'.

journal. The flora can then be presented in a simpler form for wider use. (See figs 31 & 32).

I [**8. G. verrucosum** *Huds.* in Phil. Trans. Roy. Soc., B, 56: 251 (1767); Sibth et Sm., Fl. Graec. Prodr., 1: 93 (1806); Fl. Graec., 2: 24, t. 133 (1813).

II { *Valantia aparine* L., Sp. Plant., ed. 1, 1051 (1753) nec *Galium aparine* L., Sp. Plant., ed. 1, 108 (1753).
Galium valantia Weber in Wigg., Primit. Fl. Holsat., 12 (1780); Dandy in Watsonia, 4: 48 (1957); Osorio-Tafall et Seraphim, List Vasc. Plants Cyprus, 97 (1973).
G. saccharatum All., Fl. Ped., 1: 9 (1785); Unger et Kotschy, Die Insel Cypern, 260 (1865); Boiss., Fl. Orient., 3: 67 (1875); Holmboe, Veg. Cypr., 171 (1914); Post, Fl. Pal., ed. 2, 1: 594 (1932).

Ia [TYPE: Cultivated in Chelsea Physic Garden (BM !). a

III { Spreading or sprawling annual; stems generally less than 15 cm. long, usually branched at base, sharply tetragonous, glabrous or with the angles
. . . .
mericarp through abortion; pericarp rather pale brown, coarsely and conspicuously verrucose-papillose.

HAB.: Cultivated and waste ground, roadsides, stony hillsides, shingly beaches or crevices of rocks or walls; sea-level to 5,000 ft. alt.; fl. Febr.–April.

DISTR.: Divisions 1–4, 7. Widespread in southern Europe and the Mediterranean region.

IV { 1. Fabrika Rocks near Kato Paphos, 1974, *Meikle* 4009 ! a
2. Prodhromos, 1862, *Kotschy* 849.
3. Amathus, 1964, *J. B. Suart* 157 ! a
4. Cape Greco, 1862, *Kotschy* 139A.
7. Pentadaktylos, 1880, *Sintenis & Rigo* 183 ! and, 1950, *Casey* 976 ! Kyrenia, 1932, *Syngrassides* 687 ! and, 1956, *G. E. Atherton* 389 ! 1959, *Casey* 1130 ! Klepini, 1960, *N. Macdonald* 66 ! a a a

FIG. 32 — Example from Flora of Cyprus. **I** correct species name and number; **Ia** type of correct species name; **II** synonymy; **III** description; **IV** specimen citations. Note use of **a** '!'.

GLOSSARY OF TERMS AND ABBREVIATIONS RELEVANT TO CURATION

Sometimes terms and abbreviations found both on determination labels and in floristic and revisionary works can need some explanation. Those listed below are the most frequently encountered; the list is not complete and the definitions emphasize curatorial implications. If further information is required see Jeffrey (1982), and the *latest* edition of the ICBN Examples in the list are based on the rank of species, but are equally applicable to other ranks.

aff. (*affinis*) = related to (or bordering). Placed after sp. (or sp. nov.) and before the species name (e.g. *Gardenia* sp. aff. *augusta*). This implies that the specimen does not exactly match that species but is related to it. (see fig. 33, 3). It is a more definite expression of affinity than cf./cfr. (see below).

Appr. (*approbavit*) = he (or she) confirmed. Means the same as *confirmavit.* (see fig. 33, 8).

auct. see under *sens. auct.*

cf. or **cfr.** (*confer*) = compare. This can be placed before the species name to imply that the specimen does not exactly match that species but should be compared with it. (see fig. 33, 4 and above, *aff.*).

coll. (*collegit*) = he or she gathered. Placed before the collector's name on a data label. Also an abbreviation for 'collector'.

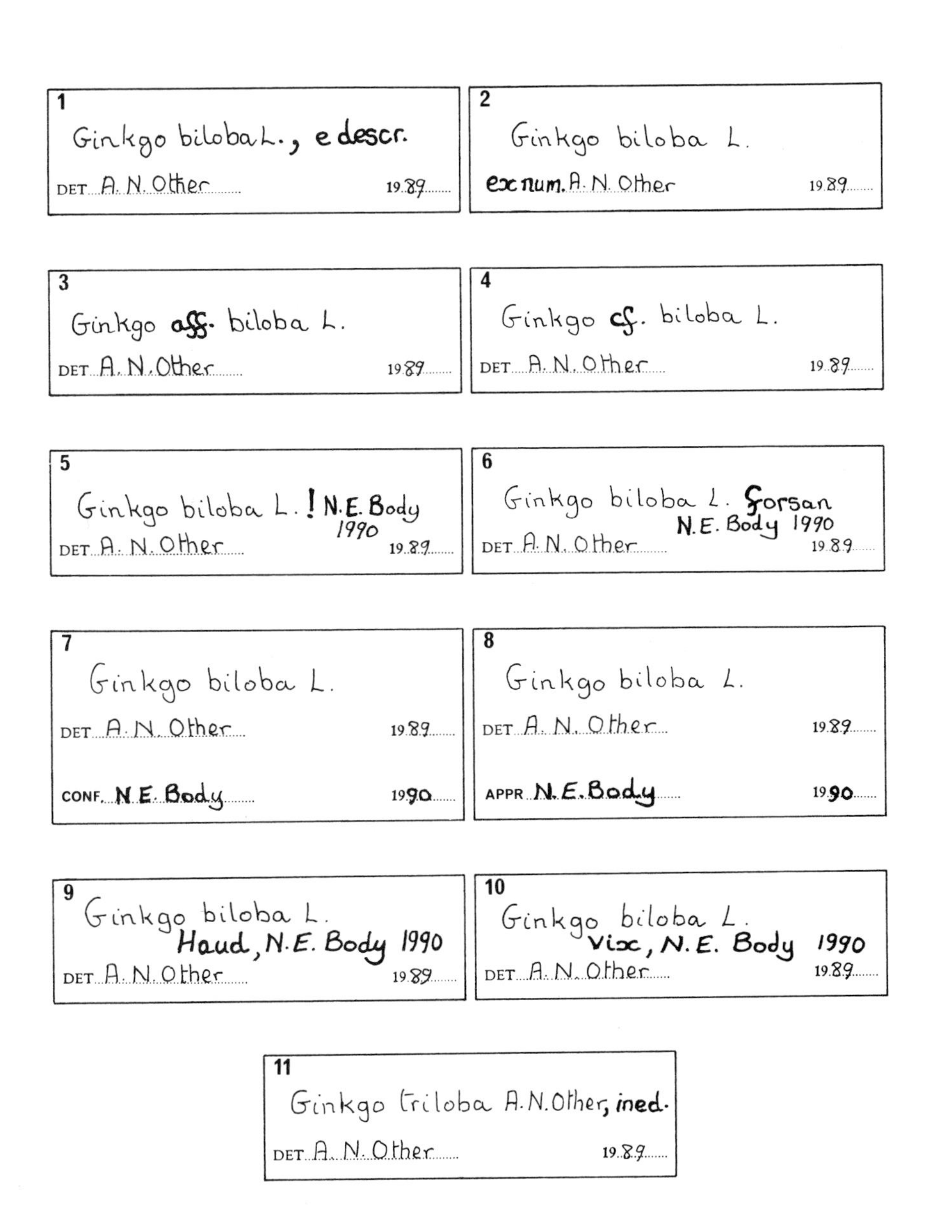

FIG. 33 — Determination labels with examples of commonly used terms or abbreviations.

comb. nov. (*combinatio nova*) = new combination. When a species is transferred from one genus to another and the same specific epithet is used, the new name is a new combination, e.g. *Beloperone scansilis* Rizzini when transferred to *Justicia* by V.A.W. Graham became **Justicia scansilis** (*Rizzini*) *V.A.W. Graham* **comb. nov.** (see fig. 30,a; double author citation under **11. Plant Names**).

comm. (*communicavit*) = he or she communicated. Used on data labels to indicate the person who sent the specimen to the herbarium; this is usually not the person who collected it.

conf. (*confirmavit*) = he or she confirmed. This is printed or written on the label followed by the name of the botanist who is agreeing that the specimen is correctly named. (see fig. 33, 7).

det. (*determinavit*) = he or she determined (or identified). This is usually printed on the label and followed by the name of the botanist who made the identification. (see fig. 33).

e or **ex descr.** (*e descriptione*) = from (or according to) the description. The specimen has been named by comparing it to a description (or by keying it out) but authentically named material with which it can be compared has not been seen. (see fig. 33, 1). Sometimes *ex char.* (= from the characters) is used instead.

e or **ex num.** (*e numero*) = from the number. The specimen has been named by tracing the collector's name and number in citation lists. This particular duplicate has not been annotated and probably not seen by the author of the revision. There is therefore always a slight possibility that owing to an error, e.g. in numbering, the specimen is not correctly named. (see fig. 33, 2).

exsiccatus = *dried*; **plantae exsiccatae** = herbarium specimens; **fungi exsiccati** = dried fungi.

- The term *exsiccata* refers to a set of herbarium specimens usually provided with printed labels; mostly accurately names sets of duplicates chosen to be uniform and representative of the taxon (e.g. '*Schedae ad Herbarium florae U.S.S.R.*').
- A *list of exsiccatae* is often published in revisions or monographs (see above, *Citations*).

f. or **fil.** (*filius*) = the son. Placed after an author's name (e.g. 'Hook.f.' is J.D. Hooker the son, while 'Hook.' is W.J. Hooker the father). For the use of '**f.**' in plant names see **11. Plant names (nomenclature)** p. 85.

forsan = perhaps. Used to indicate a doubtful determination (see fig. 33, 6).

haud = not at all. Used to disagree, usually with a determination. (see fig. 33, 9).

ined. (*ineditus*) = unpublished. Written after a name to indicate that it had not been published at that date. (see fig. 33, 11).

leg. (*legit*) = he (or she) gathered. Placed before the collector's name on a data label.

MS (*manuscriptum*) = manuscript. This indicates that the name was never effectively published; it may, for example, have just been written on a herbarium label.

nom. illegit. (*nomen illegitimum*) = illegitimate name. Used to indicate a name that has been validly published, but not in accordance with the rules of the ICBN and therefore should not be used.

nom. nov. (*nomen novum*) = new name. Used to indicate that the species has been given a new name for some nomenclatural reason. It is *not* a new species and will have previously been known by an incorrect name.

nom. nud. (*nomen nudum*) = naked name. This is a name of a taxon published without an adequate description, and is therefore not validly published.

non = not.

n.v. (*non vidi*) = I have not seen (it). This is used in the citations especially of type specimens to indicate that the author has not seen the specimen being referred to.

p.p. (*pro parte*) = in part (partly). This term follows the citation of an epithet and indicates that (in the present author's opinion) more than one taxon was included under this name by the author cited, i.e. the specimens cited in the previous work included some belonging to the present species and some belonging to a different species. (see fig. 30, d).

pro parte is sometimes also used to indicate mixed gatherings (e.g. *Smith* 123 p.p.).

quoad = as to the (as regards or with respect to). Often found following citations of an epithet when **pro parte** has been used. The specimens listed after *quoad* are the ones that belong to the taxon in question and should be relabelled and placed with it. (see fig. 30, e).

sched. (*scheda*) = label. Used in literature to indicate information taken from a herbarium label (*in sched.* = on a herbarium label). See also '*exsiccata*' above.

sens. (*sensu*) = in the sense of. If in literature this term is placed between a specific epithet and an author's name it implies that that author used the name incorrectly, perhaps as the result of a misidentification. Such names are usually listed with synonyms but they are not true synonyms. The specific name as used by the original author will belong to a different taxon. (see fig. 30, c).

sens. auct. or **auctt.** (*sensu auctorum*) = in the sense of some or many authors. The use of this term is similar to that given above except, that numerous authors have habitually used the epithet incorrectly.

s.l. or **sens. lat.** (*sensu lato*) = in the broad sense. This term often follows a specific name as an indication that it is being used in a wide taxonomic sense – i.e. the species may include several varieties or forms or even closely related species that may be difficult to distinguish.

s.n. (*sine numero*) = without a number. This follows the collector's name to indicate that the specimen was unnumbered.

sp. = species (singular); **spp.** = species (plural).

s.s. or **sens. str.** (*sensu stricto*) = in the strict sense. This term often follows a specific name as an indication that it is being used in a narrow sense – i.e. not including other possible species which are included when the name is used in the broad sense; or if the species includes varieties, only the typical variety is being referred to, i.e. where the varietal epithet is the same as the specific epithet.

ssp. or **subsp.** = subspecies.

stat. nov. (*status novus*) = new status. Used in literature to indicate that the author has altered the rank of a validly described taxon – e.g. a species may be reduced to the rank of subspecies or a variety may be raised to the rank of species.

syn. (or **synon.**) (*synonymum*) = synonym. When two or more names have been used for the same taxon, the first name to be published has priority (see ICBN) and must be used as the name of the taxon. Any later names are regarded as synonyms.

syn. nov. (*synonymum novum*) = new synonym. Used to indicate that the author believes it to be the first time that the name has been treated as a synonym of the present taxon. (see fig. 30, b).

teste = according to or by the witness of. Sometimes used instead of '*det.*', see above.

ubi? = where?. Used following the citation of e.g. a type, to indicate that the present location of the specimen is not known.

var. = variety.

vix = scarcely. Used, e.g., on determinavit slips and usually initialled. It is placed against a previous determination to indicate that the second botanist disagrees with it (see fig. 33, 10).

! When used on a determination label with initials (or signature) placed against a previous determination: indicates that the second botanist agrees with the previous determination. (see fig. 33, 5).

When used in literature following the collector's name and number: This indicates that the author has seen the specimen (see figs 31, a & 32, a).

† = This symbol is used in literature, e.g. after citation of a type specimen to indicate that it has been destroyed. (see fig. 31, b).

[ ] = Square brackets are often used in lists of synonyms to indicate an erroneous use of a name, rather than a true synonym (see *sens.* or *sens. auct.*) or an illegitimate name (e.g. homonym). (see fig. 30, f).

♀ = female

♂ = male

⚥ = hermaphrodite, i.e. male and female

∞ = (infinity). Numerous

§ = section of a genus

REFERENCES

Fosberg & Sachet: 78–81 (1965)
Gleason (1933)
Jeffrey (1982)
Stearn (1983)

24. THE DISSECTION OF FLORAL ORGANS AND PRESERVATION OF THE RESULTS

The principal reason for dissecting a flower is to investigate its structure. This is important both for identification and classification.

Unless very many flowers are present on a specimen the dissections should be carefully arranged and preserved for examination by later workers. It is important to keep all the parts of a flower even if some have been damaged during dissection.

Since some information can be lost in the preserved dissection it is important to record it and, if possible, prepare a sketch. Always record:

- The number of parts and their arrangement, i.e. petals opposite or alternate to the sepals.
- The type of aestivation of the sepals and petals: whether and how they overlap. This can often only be observed in buds.
- The measurements of the various organs, while still fresh or moist.

Equipment

Fine mounted needles
Fine forceps
Sharp scalpel with fine point
Small paint brush
Watch-glass or petri dish

Choice of material

Living: It is usually much easier to dissect a living flower than a preserved one. Where possible, collectors should be encouraged to prepare limited dissections in the field, e.g. large tubular corollas can be slit along their length and opened out before pressing. Some polypetalous irregular flowers, e.g. *Impatiens* or papilionoid legumes, can easily have their petals separated before pressing.

Herbarium: Herbarium specimens are valuable and it is important to be economical with material as most sheets do not have an abundant supply of flowers. Take care over choosing which flowers to dissect. In some groups, e.g. *Asclepiadaceae,* floral structures can become distorted when pressed, and may not regain their original shape on boiling. In such groups try to select flowers that have been protected from excess pressure by their proximity to some other organ (a branch, for example).

Sometimes large-flowered taxa have only one flower on the sheet and its removal would ruin the specimen. It is possible to carry out limited *in situ* dissection in some cases. If the corolla is tubular (e.g. many Bignoniaceae, *Datura* (Solanaceae), or *Rothmannia* (Rubiaceae)) a portion can be softened by locally applying a wetting agent (see below, *Using a wetting agent*) and a flap cut and folded back to reveal the internal structures. The flap should be replaced and the flower blotted and allowed to dry under light pressure. Alternatively, the flap may be removed and retained in a paper capsule; the flower should then be protected by a 'window' (see **9. Mounting herbarium specimens**). It is also possible to examine the ovary of some taxa (e.g. *Spathodea*) by cutting a similar flap in the calyx; the wetting agent will require about half an hour to work on the more resistant calyx tissues.

Spirit: Floral structures are often easy to observe in spirit-preserved material, especially where 3-dimensional structures are important. If there is spirit material supplementing the herbarium specimen, this may avoid the need to dissect dried flowers.

Most liquid preservative mixtures contain 1% glycerol, which helps to prevent the plant parts becoming too brittle. It also prevents the flower from drying up during dissection and, in some cases, the flower may be left in a soft condition when the mixture has evaporated. If the specimen was preserved in a fixative solution containing formalin, it is advisable before dissection to rinse it in dilute spirit.

Preparing material for dissection

Dried flowers must first be softened before they can be dissected. There are several methods that can be used:

- *Boiling in water*. Place the flower in gently boiling water until it is soft. This may be only ½ minute for small or delicate flowers – large or fleshy flowers take longer to soften. Remove the flower from the boiling water with a pair of forceps and place in a shallow dish or watch-glass containing water.

- *Using a wetting agent*. Pohl's solution is a softening and wetting agent (see Pohl 1965). Other wetting agents are available, e.g. 'Libsorb' (from Allied Colloids Ltd). Very dilute detergent can also be used. Place the flower to be softened in a watch-glass with several drops of the solution and leave for a few minutes.

  For delicate flowers this method is better than boiling. It should always be used when, for example, delicate corollas are combined with a tough calyx and/or ovary.

- *Using ammonia*. For dried fleshy flowers which do not produce good results with boiling water (e.g. *Orchidaceae*, *Asclepiadaceae*) the use of ammonia is recommended. This method usually restores the 3-dimensional form of the flower. Place the flower in a strong ammonia solution in a fume-cupboard and leave overnight. The following day, remove the flower and wash it in several changes of water. Place the flower in a shallow watch-glass with some water.

Dissecting the flower

Softened flowers must be kept moist and can be dissected covered in water in a watch-glass or shallow dish, or on a slide or tile. Spirit preserved flowers will usually remain moist, but a few drops of spirit may be added if needed. Delicate material may be dissected more easily on a firm surface.

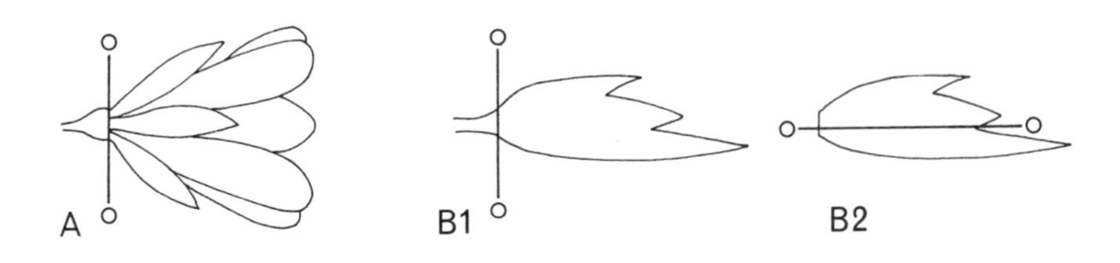

FIG. 34 — Position of initial cuts on the upper surface for separating floral parts. **A** flowers with free parts; **B** flowers with fused parts, **1** first cut and **2** second cut.

Flowers with free parts:

If the perianth segments will not easily separate, first cut at the extreme base of the sepals (Fig. 34,A). The sepals should then fall off, and can be moved carefully to expose the base of the petals. The petals may also fall off but if still attached each should be carefully pulled away with forceps and cut at the base if necessary. The stamens will now be exposed and can be removed in a similar manner, leaving the ovary, style and stigma. It is usually only necessary to remove the organs on the side facing the observer. In some cases the ovary may require further dissection if the placentation type must be determined and the ovules counted.

Flowers with fused parts:

First detach the corolla-tube and split along one side between the lobes using a mounted needle or small-bladed scalpel, then open the corolla-tube out. The calyx-tube should first be cut at the base (fig. 34,B1) and then split along one side, in the same way as the corolla-tube, before removing it (fig. 34,B2). The remaining organs can be treated in the same way as flowers with free parts. However, in many families with a fused corolla the anthers are attached inside the corolla and will need no further treatment.

Arranging and preserving the dissection

The dissected parts can be arranged using the brush. They should be measured and if necessary drawn while still moist. It is important to observe them as they

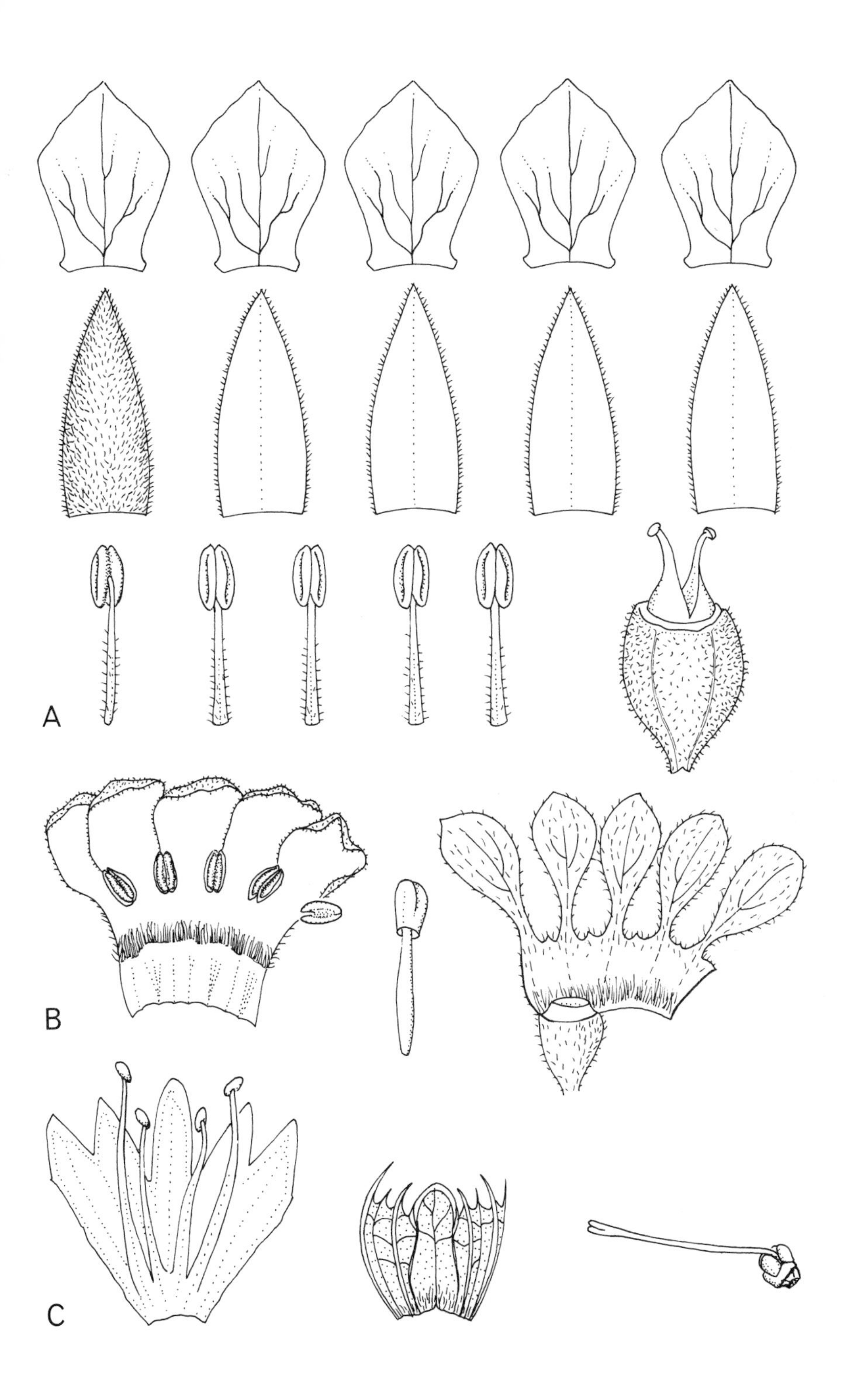

FIG. 35 — Suggested arrangements of dissections. **A** regular flower with free perianth and inferior ovary; **B** regular flower with tubular perianth and inferior ovary; **C** irregular flower with tubular perianth and superior ovary.

continue to dry, since fine hairs and other structural details will often not show up under a thin film of water.

When the sepals and petals of a regular flower with free parts are similar in shape and size they should be arranged to show both sides. For flowers with fused parts, the spread-out calyx-tube and corolla-tube should be placed so that the inside is displayed to show any anthers, hairy tufts or glands (see fig. 35). Note that the parts must be placed upside-down (i.e. face-down) since they become reversed on the mounted dissection. Any excess moisture should be removed with blotting or tissue paper. A piece of gummed linen cloth-tape or white paper is then placed over the arranged dissection, gently pressed down and carefully lifted, making sure that the dissected parts are stuck to it.

If the material is limited and it is necessary to re-examine the dissection in the future, moisten the gummed paper, ease off the organs and place in a drop of water containing a little detergent under the microscope.

Alternatively, the dissection can be floated onto a strip of strong (5 mil) transparent polyester film, the parts arranged, the excess water blotted off and a second polyester strip placed on top. The two strips must be glued in place with a dab of suitable clear glue at each corner. The dissection will dry and will be held firmly in place.

Some material (e.g. *Leguminosae*) contains sufficient natural adhesive to stick to plastic or mica (see below, *Particular groups*).

Dissections should not be sandwiched between strips of 'Sellotape' or similar tape; although this originally makes an attractive mount, the adhesive fails with age and the tape discolours and becomes brittle. Archival quality tapes have recently become available but no data are known as to their effectiveness.

Permanent mounts of dissected herbarium material can be made by arranging the dissected parts on a glass slide, removing excess water with blotting paper and allowing to dry slightly. A few drops of sodium silicate (water-glass) are then dropped onto the slide and the preparation sealed with a cover-slip (Quisumbing 1931).

Some flowers have fleshy parts and it may be better to preserve the dissection as part of the spirit collection. The dissected flowers should be placed in a small tube or bottle filled with sufficient spirit mixture to cover the flower. Remember to place a copy of the information from the herbarium sheet with the tube and to cross-reference the herbarium and spirit collections.

If a dissection is made from a specimen preserved in spirit, it should be placed in a smaller sized tube which will fit inside the original container and a cross-reference placed in the herbarium if necessary.

Particular groups:

Leguminosae – Papilionoideae

Detach calyx as in flowers with fused parts. Remove the petals by severing them at the base with a mounted needle. The staminal sheath and the ovary can be removed in a similar manner.

Place the corner of a mica sheet in the watch-glass and slide the floral organs onto the sheet. Arrange them as in fig. 36 – by convention the claws face left. N.B. if the keel has separate claws, open it out. Pat dry with good quality blotting paper. Remember to label the mica sheet with the collector's name and number, place the mica sheet in a polyester packet and put it in a paper capsule attached to the herbarium sheet.

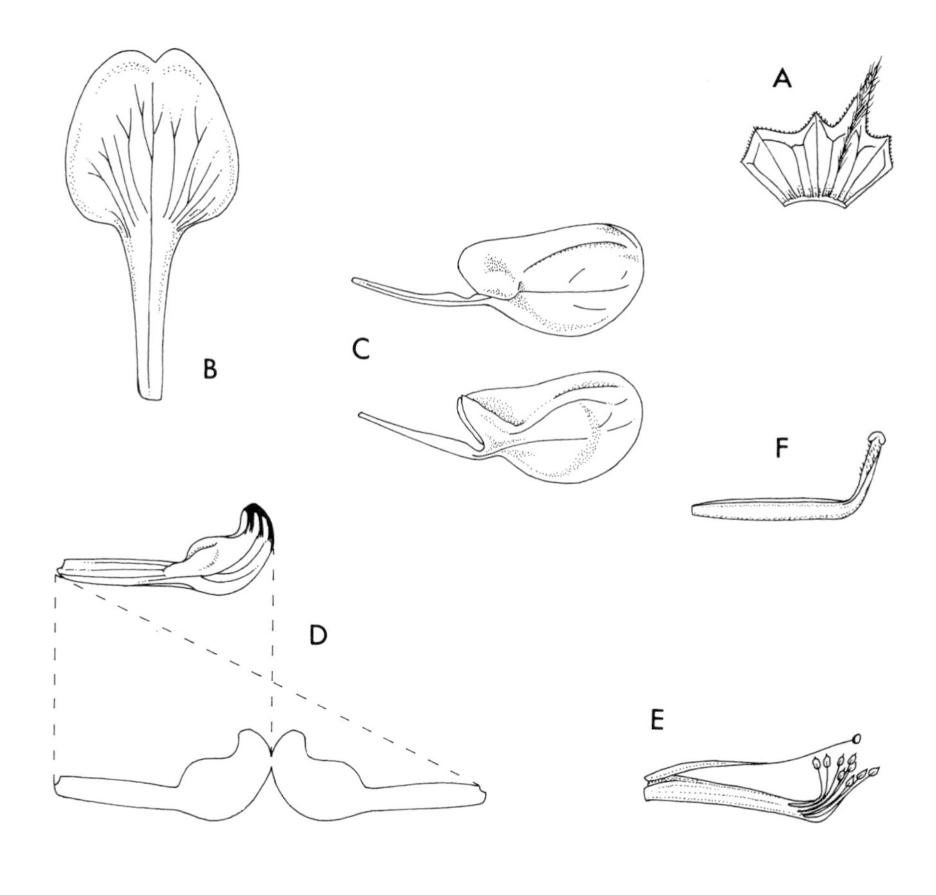

FIG. 36 — Lay-out of Papilionoideae dissection. **A** calyx; **B** standard; **C** wings; **D** keel; **E** staminal tube with free stamen; **F** ovary.

The attachment of the column to the labellum (lip) is important in the Orchidaceae, so these two organs are normally left together (fig. 37). It may be necessary to slice the labellum longitudinally to observe the morphology of any canals or chambers. The remaining perianth parts should be detached. The dissected parts should either be dried and returned to a paper capsule on the herbarium sheet, or placed in a labelled jar with spirit.

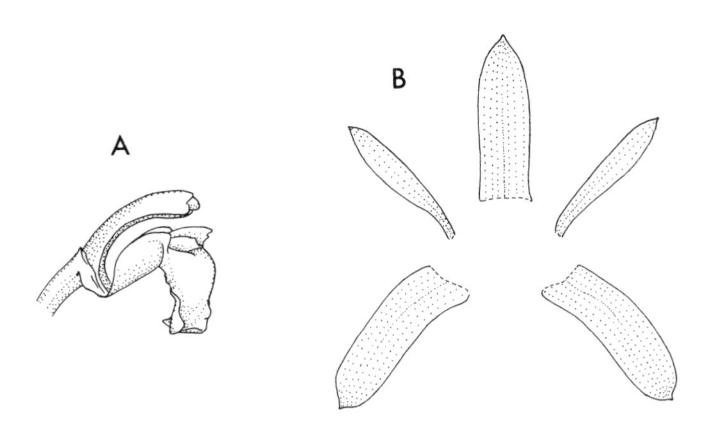

FIG. 37 — Lay-out of Orchidaceae dissection. **A** column and labellum; **B** remaining perianth segments.

REFERENCES

Pohl (1965)
Quisumbing (1931)

25. ILLUSTRATION

Many small herbaria may be unable to afford to employ artists to produce botanical illustrations, maps and diagrams for their own archives or to accompany publications. The ability to draw is a skill many technicians may find worth developing. Although talent certainly helps, most people can learn to produce adequate illustrations; practice is of paramount importance.

The artist will usually work under the supervision of a botanist who will explain, where necessary, the morphology and what is important. In any event, it is essential that the artist understands the morphology of the plants being drawn and produces accurate results. Skill in dissecting under a binocular microscope is a distinct advantage (see **24. The dissection of floral organs and preservation of the results**).

Black and white line drawings are most commonly used to accompany botanical papers and flora accounts. Such drawings need only be simple; complex drawings with a lot of fine detail often do not convey any extra relevant information. Colour plates are very expensive to reproduce and are only used in a few productions. Illustrations indicating flower colour can be usefully added to the records in the herbarium, but colour techniques are not dealt with here.

Drawings must be made precisely to scale and this recorded (usually on the reverse side). Drawings of the habit (except for very small or very large subjects) are conveniently done at natural size (× 1), but details of flowers and seeds for example are usually enlarged. If the illustration is for reproduction bear in mind the final size of the printed plate and choose a measurement which will scale down to a sensible figure: e.g. if the drawing is to be finally reduced by ½ a detail drawn 6 times larger than life (× 6) will appear in the publication as × 3; or if the plate is to be reduced by ⅓ a detail at × 6 will be reduced to × 4. The usual

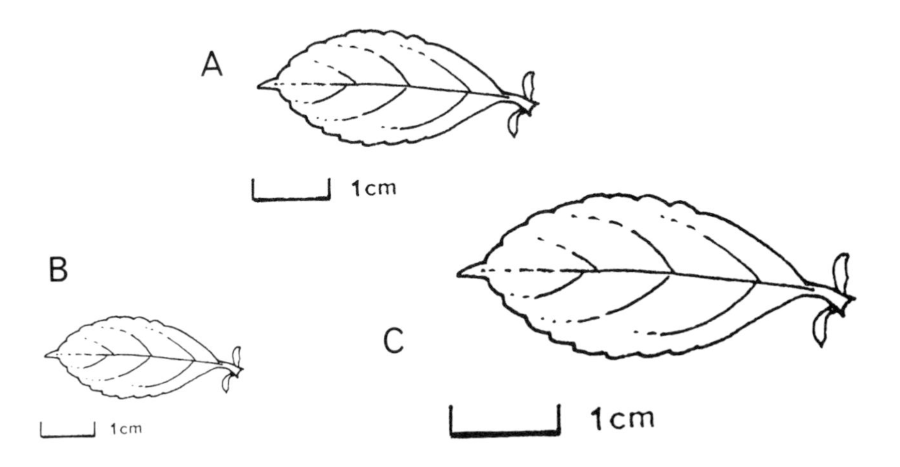

FIG. 38 — The use of a scale-line. **A** natural size (× 1); **B** reduced; **C** enlarged.

reduction for most publications is $1/3$, i.e. reproduced at $2/3$ the size of the drawing. Alternatively a scale-line can be added to each detail (useful if the final reduction is not known) (see fig. 38). There are several aids to scale-drawings:

- Use **graph paper** – helpful for detached items of moderate size such as a fruit. Transfer to final drawing with tracing paper. (fig. 39, A).

- Draw a **frame** or **grid** of the required size and scale (e.g. 3 times larger than item) with a ruler and set square – use for details that are glued to the sheet. A natural sized grid (× 1) on a transparent sheet can be placed over the item to act as a guide if necessary. The best way to make such a grid is to draw it in pen and ink on paper, then photocopy it onto an acetate sheet (the kind used for overhead projectors). If this is not possible it may be drawn directly on tracing paper or transparent film. (fig. 39, B).

- Use a **graticule** in the eye-piece of your microscope or a lens with a built-in scale (in 1 mm. and 0.1 mm. divisions) to measure directly microscopic details. Then draw within a constructed frame or grid as above.

- **Proportional dividers** are useful, but can be rather costly and are not essential (not useful for microscopic details).

- There are many more sophisticated methods (e.g. pantograph, camera lucida, reducing and enlarging photocopiers and photographic apparatus) but these rely on expensive equipment and will not be considered here. In practice most competent botanical artists prefer not to use them.

Some understanding of 3-dimensional perspective can help in the achievement of a realistic drawing; it can be a difficult skill to acquire but is not usually essential for the conveyance of accurate information. Herbarium specimens have been artificially distorted into two dimensions and are best portrayed in a 2-dimensional way unless drawn by a skilful artist. Practise drawing from live material as much as possible. Remember a 'stiff' but accurate drawing is scientifically preferable to a 'lively' but inaccurate one.

The simplest form of drawing is **line drawing** and this is quite adequate to convey basic information such as leaf-shape or, if skillfully done, more complex structures giving a 3-dimensional effect (see fig. 40). Usually, however, more complex forms need some kind of shading to help explain their shape, form and surface texture. There are two main types of shading used for black and white illustrations: stipple and hatching.

- Stippling – this is the easiest method for the beginner to handle as it builds up the shading slowly. Stippling is best done with a stylus pen (e.g. 'Rapidograph'), but a rigid pen-nib will do. The dots should be far apart for light areas and become closer together for darker areas (see figs. 41, A & 42). If the drawing is to be reduced, remember – dots that are too fine will tend to be lost, while dots that are too close together will merge, over-darkening the

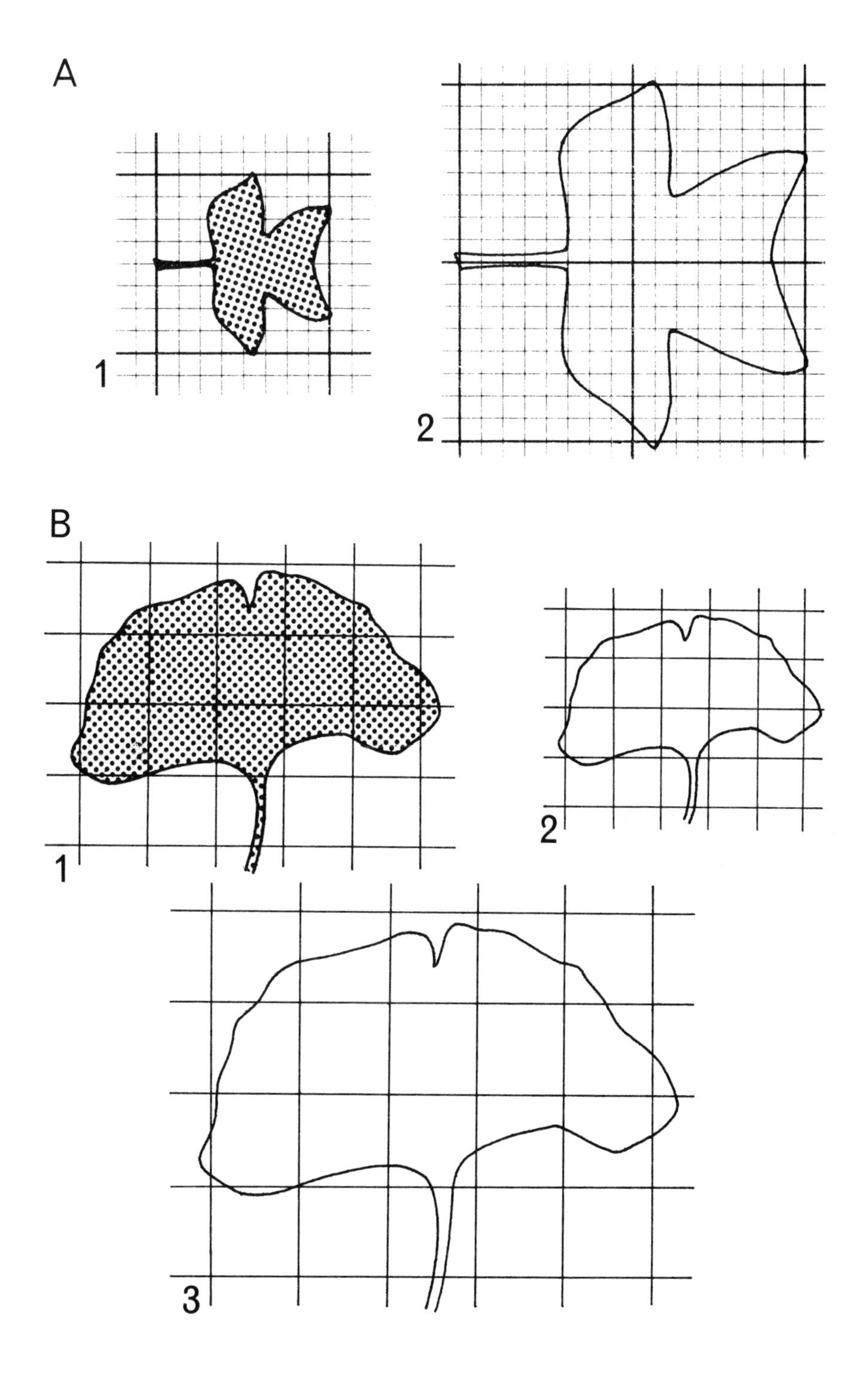

FIG. 39 — Methods of drawing to scale. **A** by use of graph paper, **1** item placed on graph paper, **2** drawn twice original size (× 2), i.e. 2 × 2 squares for every one; **B** by grid drawn to required size, **1** natural size (× 1): 1 cm², **2** ⅓ smaller (× ⅔): 0.66 cm²; **3** ⅓ larger (× 1 ⅓): 1.33 cm².

shadow. The main drawback with stipple shading is that it can be confused with surface texture or short fine hairs (puberulous indumentum), and care must be taken to prevent this.

- Hatching – takes more skill than stippling but once mastered can be done in less time. Generally the hatching lines should run with the contours of the subject and become closer together as the tone darkens; the darkest tones can be **cross-hatched.** While learning it is best to sketch the hatching lines in pencil first; the best results will be obtained using a pen with a flexible nib. The main disadvantages with hatching are that it can give a ribbed appearance (especially to stems) and does not allow such fine detail as stippling. Remember that if a drawing is to be greatly reduced, hatching will obscure detail in the printed version. (See figs. 41, B, C & 43).

Professional effects can be otained by combining both stippling and hatching in the same drawings but this is not recommended for the beginner.

Points to remember:

- First decide on the size of the drawing and measure the frame. Check that this will reduce to the required size of the published plate.

- In general the botanist will instruct you as to which botanical details are required and of any special features to be depicted.

- Remember to show both sides of the leaf in the habit drawing.

- Do not draw irrelevant details, particular to the individual plant rather than the taxon – e.g. holes caused by insect damage.

- Draw the details on separate pieces of paper and trace them; arrange the tracings in your frame so that they form a pleasing balanced composition; transfer the tracings to the plate and firm up the pencil lines.

- Record on the back of the plate:

  a. The name of the taxon.
  b. The scale of each item (before and after reduction).
  c. The collector's name and number of the specimens from which the drawings were made.

- To draw a symmetrical object, first draw half, then trace and reverse the tracing adjacent to the first half. This method is particularly useful for diagramatic work, but natural shapes, it should be noted, are rarely 100% symmetrical.

- Draw lightly in pencil then ask a botanist to check the details before inking-in.

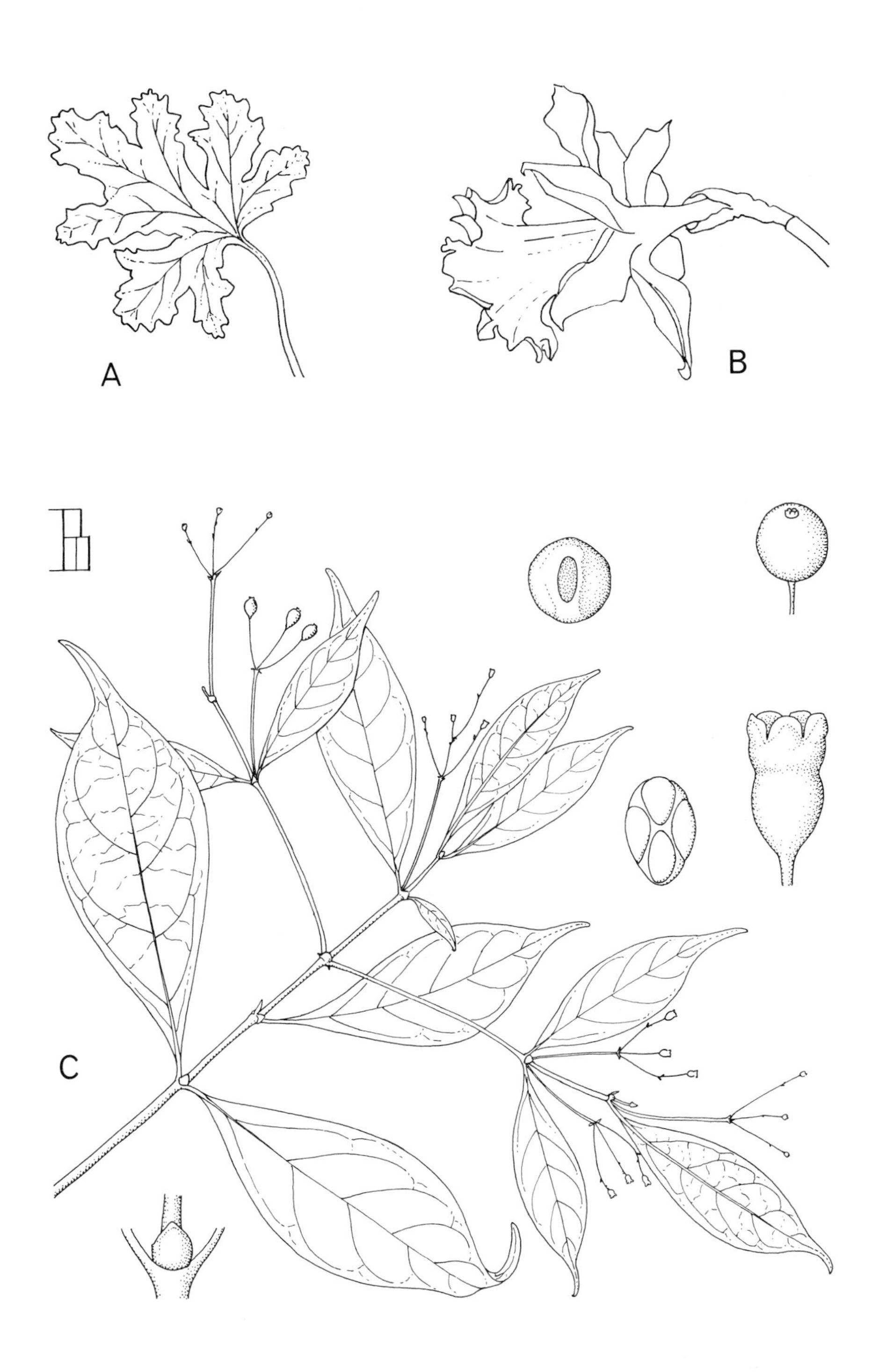

FIG. 40 — Examples of line drawing. **A** simple outline of leaf, *Geranium* sp.; **B** flower of *Narcissus pseudonarcissus*; **C** habit drawing of *Tarenna uzungwaensis*, with shaded details.

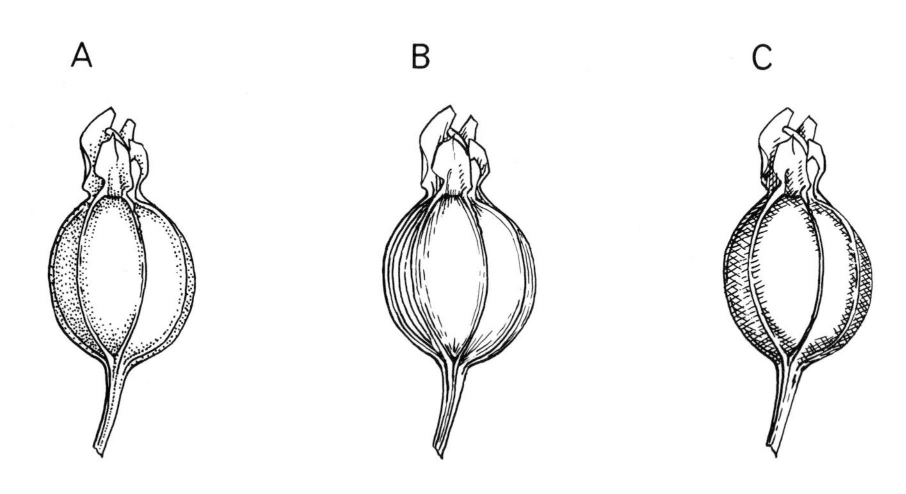

FIG. 41 — Examples of shading. **A** stippling; **B** hatching; **C** cross-hatching.

— Ink in the outlines boldly, except where a broken or dotted line may be needed to indicate a densely hairy subject.

— Always shade on the same side. Imagine the light shining from only one direction, taking account of leaves shading each other, etc.

— Take care over details of indumentum (hairs) and surface; do not draw them too finely or too coarsely. (see fig. 44, A & B).

— In general do not try to represent colour by tone. However, in some patterned subjects (e.g. spotted petals or blotchy seeds) it may be necessary to indicate the coloured areas; try to make the shading distinguish both pattern and form. (see fig. 44, C).

— Try to visualize the drawing after reduction. It is helpful to use a reducing lens if one is available. Very fine lines and dots tend to vanish. Shading will tend to darken, especially the darker tones. If in doubt, err slightly on the bolder side and make the shading lighter.

— Lettering or numbering – this is best done with 'Letraset' or a similar product. (See fig. 45).

— Use 'Zip-a-tone' or similar product for shading in areas of even tone in diagrammatic and map work. (See fig. 45).

— Removal of errors:

a. If the drawing is on good quality Bristol board or drafting film, small errors can be removed by scratching with a sharp blade and then smoothing the area over.

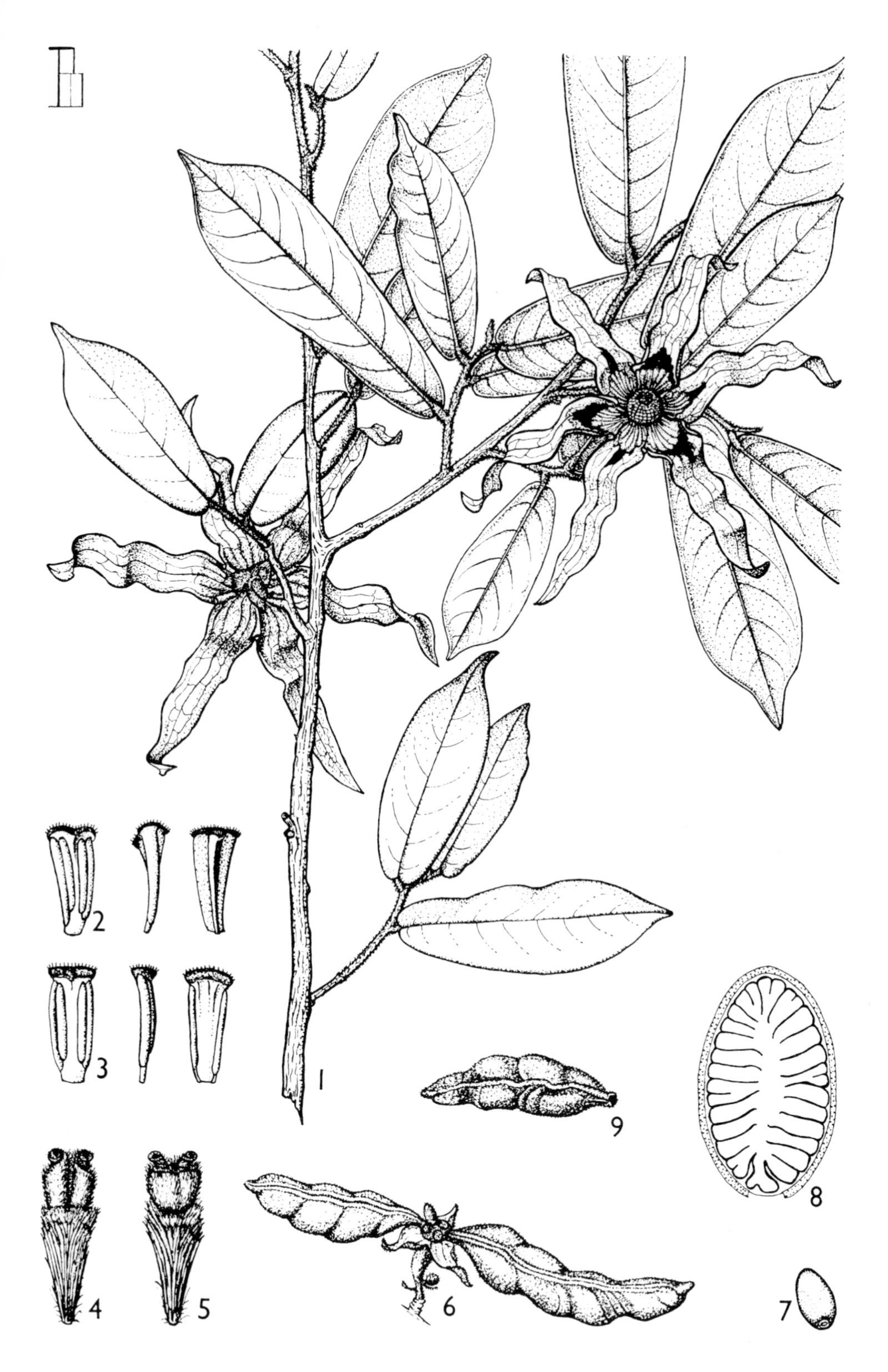

FIG. 42 — Example of a drawing shaded by stippling (*Asteranthe asterias* (Annonaceae), by Diane Bridson).

FIG. 43 — Example of a drawing shaded by hatching (*Multidentia concrescens* (Rubiaceae), by Stella Ross-Craig).

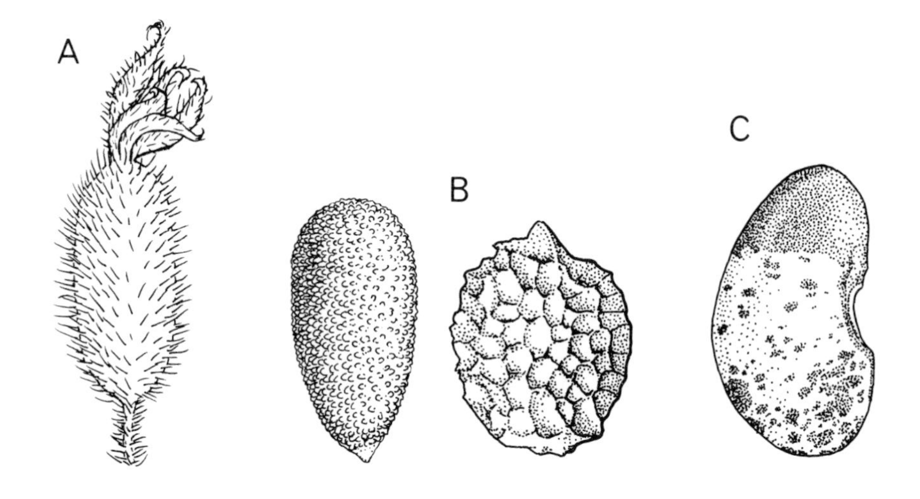

FIG. 44 — Examples of details. **A** indumentum; **B** texture; **C** colour (or pattern) combined with form.

b. Errors can be painted out with process white, and the detail redrawn - after the white paint is *completely* dry. 'Tipp-Ex' can be used but it is important to remember that it may deteriorate with age.

c. Redraw the detail on a separate small piece of paper and glue to the plate. The edges of the inset will not show when the plate is reproduced.

Materials

- Inexpensive sketching paper (for initial sketches).
- Bristol board, or the best quality drawing paper obtainable, the smoother the surface the better. Choose white, not cream or tinted.
- Pencils – hard pencils (6H) are best for drawing fine detail, but these need to be kept very sharp (use emery paper). Also keep a selection of softer pencils (4B–HB or F) for rough sketching. Propelling pencils such as 'Pentel' with very fine leads (0.3 or 0.5 mm.) are excellent and save time as they do not need constant sharpening.
- A good soft eraser (some artists prefer art gum).
- Tracing paper.

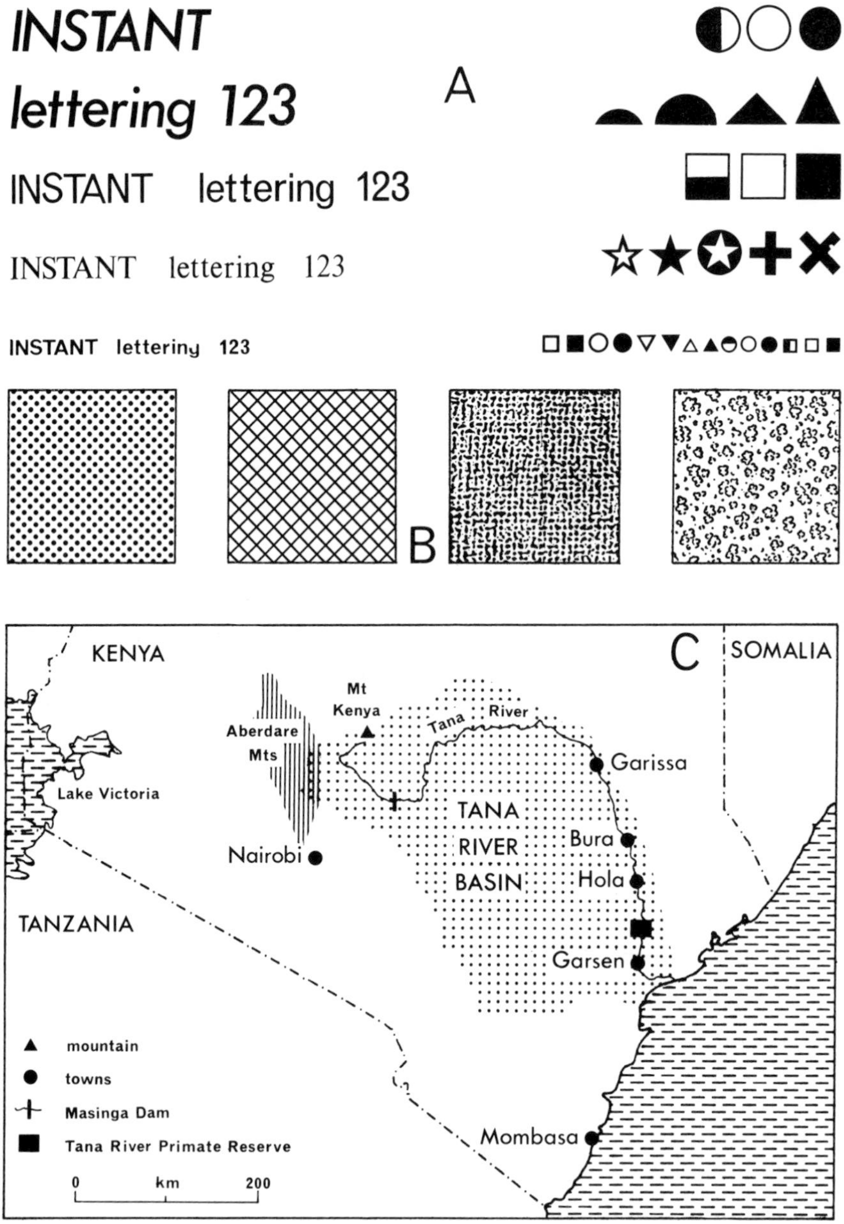

INSTANT
lettering 123

A

INSTANT lettering 123

INSTANT lettering 123

INSTANT lettering 123

B

C

KENYA

SOMALIA

Mt Kenya

Aberdare Mts

Tana River

Garissa

Lake Victoria

TANA RIVER BASIN

Bura

Hola

Nairobi

TANZANIA

Garsen

mountain

towns

Masinga Dam

Tana River Primate Reserve

Mombasa

0 km 200

FIG. 45 — Instant lettering and tone. **A** letters, numbers and symbols, × 1; **B** tones, × 1; **C** map using symbols, letters and tones.

- Frosted acetate film (drafting film) such as 'Permatrace' – use for maps and diagrams. Take care, as normal ink takes longer to dry than on paper. Mistakes can be very easily scratched away.

- Pens – choose a flexible nib for variable lines (hatching and outlines) or a rigid nib for other details (stippling or ruled lines).

- Stylus pens (e.g. 'Rapidograph'). These give a line of constant thickness, are excellent for stipple work and convenient to handle. However, they are expensive and can be easily damaged (especially the fine sizes). Always follow the manufacturer's instructions; never allow the ink to dry on the nibs.

  Use: 0.3–0.5 mm. nibs for general outlines.
  0.2–0.25 mm. nibs for finer lines (e.g. nerves of a leaf).
  0.1–0.12 mm. nib only for very fine detail such as hairs.

  Larger sizes are useful for maps and diagrammatic work requiring a bold outline.

- Ruling pen and compass set with interchangeable, adjustable nib. This is useful for diagrammatic work, especially if a stylus pen is not available.

- Ink – use waterproof Indian ink with dip-pens but *always* use special ink in stylus pens.

- A sharp blade for pencil sharpening and correcting minor errors.

- If products such as 'Letraset' and 'Zip-a-tone' are available they will give a professional finish to map and diagrammatic work (see fig. 45).

- Ruler, preferably with 0.5 mm. divisions, and a set square.

REFERENCES

Dalby & Dalby (1980)
Holmgren & Angel (1986)
West (1983)
Wood (1979)
Zweifel (1961)

26. THE PHOTOGRAPHIC COPYING OF HERBARIUM SHEETS

There are four basic systems used to copy herbarium sheets, and the choice of system will depend upon the use of the resultant image and overall resources.

Colour transparencies:

The simplest method is to use a 35 mm. single-lens reflex (SLR) camera on a stand and produce 35 mm. transparencies onto colour reversal film. The resulting transparency can be of very high quality and should be viewed using a projector or viewer; or they can be scanned by printer for publication. Larger format cameras can be used, with resultant increase in cost per slide but improvement in image quality.

Black and white film:

The sheets may be copied onto perforated black and white negative film using a 35 mm. SLR camera. The resulting negatives are printed for viewing or publication, but for this, darkroom facilities are required. Larger format cameras can be used, but again the cost increases, though the quality of the final print is improved.

Microfilm:

If large quantities have to be copied, then a microfilm camera can be used. This takes longer lengths of unperforated 35 mm. film, giving a larger image area. The images can be viewed on a viewer. More information will be obtained from the herbarium sheets if they are filmed in colour, but black and white microfilm, which is much cheaper, is often adequate.

'Cibachrome' prints:

If high quality A4 or A3 size colour prints are required of the sheets, then the best method is for them to be copied onto 'Cibachrome Copy' material, but the initial outlay for the camera and processor is very high. Both material and equipment are manufactured by Ilford Ltd. and distributed via a network of worldwide selling agencies.

See **14. Collections of illustrations and photographs**, *Photographic records of herbarium specimens* and p. 82, *Photocopying specimens*.

REFERENCES

Womersley: 101–104 (1981)

27. COLLECTORS, ITINERARIES, MAPS AND GAZETTEERS

Information about the distribution of a species is derived from the localities given on individual specimens. It is therefore often necessary to find out exactly where a particular locality is situated, and a suitable collection of maps, gazetteers and atlases should be assembled for this purpose. These may also be useful for checking the altitude of a particular collection or the habitat by reference to vegetation maps.

Travel books and articles are invaluable, especially where the author was a collector and details of routes with dates and localities are given. Biographies and bibliographies can also be useful sources of information concerning botanical collectors and their itineraries: Stafleu & Cowan (see references below under '**General**' and **22. Essential Taxonomic Literature**) is the best initial bibliographical source. In some cases it may be necessary to compile an itinerary if one is not already available; this can be extracted from the collector's note-books in conjunction with maps and gazetteers of the area. If notebooks are not available, the specimens can be traced from a determination list and the data sampled from different number-blocks until the itinerary can be drawn up from the localities and dates given on the data labels.

It is worth noting that abbreviated dates can be a cause of confusion. The usual European sequence for numerically abbreviating dates is: day, month, year – i.e. 13.10.91 = 13th October 1991, but the sequence used in America is: month, day, year – i.e. 10.13.91 = 13th October 1991. When the day is less than 13, this can be ambiguous – i.e. 2.7.91 could be either 2nd July 1991 or 7th February 1991; it is necessary, therefore, to know if the usage is European or American. Abbreviated dates as indicated on some historical specimens can be misleading, e.g. 7bre, 8bre and 9bre (or 7ber, 8ber and 9ber) have sometimes been used for September, October and November, and not, as may be supposed, for July, August and September.

A good modern set of maps and gazetteers is essential for checking the longitude and latitude or grid reference of a locality before finalizing the data labels for new collections. With the increasing use of computers for both record keeping and the production of distribution maps, accurate information on the coordinates is essential. It is important to note that longitude and latitude shown on old maps and sketch maps may not be accurate and should therefore be checked against modern maps. Be especially careful of old French maps (up to c. 1880) in which the longitude was based on the Paris rather than the Greenwich meridian.

In addition to a good set of modern maps and gazetteers every effort should be made to acquire and/or keep the following:

- Old maps, atlases and books. These are very useful for localizing old specimens collected in places where the names or boundaries have changed. Check the date of publication and, if possible, the source and date of compilation of old maps. Never throw away such items; look for them in second-hand shops. Some may be valuable collectors' items but others may be inexpensive and worth purchasing.

- Maps annotated by collectors. Even if worn and dirty, these are most important.

- Sketch-maps drawn by collectors. If older maps are in the form of blueprints or other types of reproductions with a tendency to fade with age, a good photocopy should be taken to provide a permanent record.

- Maps produced for special local projects or uses — especially maps of forest or other nature reserves.

- Maps produced to show special features, especially vegetation maps, land-use maps and geological maps.

- Collectors' notebooks (vital for the compilation of itineraries).

- Determination lists (useful for the compilation of itineraries).

- Letters from collectors offering any explanatory notes about their itineraries.

Maps should be stored flat wherever possible; continual folding and unfolding will weaken them rapidly. Other smaller items such as sketch-maps and letters from collectors can be filed in box-files, arranged by collector and/or geographical area.

Difficulty can sometimes be experienced in finding names in gazetters or on maps. The following possibilities should be borne in mind:

— Is the word really a locality? Older data labels are often ambiguously formatted and confusion between geographical and vernacular plant names is frequent. Confusion can also occur with other kinds of unfamiliar words, especially with locally used habitat terms, e.g. maquis, fadama or with terms used to indicate 'the same as the data on the previous label', e.g. ditto, ebenda, ibid.

— Has the locality name been changed? Compare the date of the gazetteer or map with the date of collection of the specimen. If possible consult an earlier publication. It is a good idea to keep an index of old place names against their modern equivalents; this can save time in the future. Newly named national parks or forest reserves, for example, may not appear on the most up-to-date map for the area. Check the date on which the park or reserve was formally designated or gazetted, if at all possible.

— Has the boundary been moved? If a map or gazetteer devoted to one country is being used, try also one from the neighbouring country.

- Is a variant spelling of the name possible? There may be old and modern spellings of a name, e.g. Aethiopia and Ethiopia. Spelling differences due to transliteration in different languages, e.g., Mozambique–English and Moçambique–Portuguese or translation, e.g. Ivory Coast–English and Côte d'Ivoire–French. Languages normally written in scripts other than the Roman alphabet, e.g. Russian, Chinese, often result in numerous spelling variants when transliterated into the Roman alphabet and this should be anticipated when consulting gazetteers. In some parts of the world, e.g. Africa and China, 'l' and 'r' are often confused and should therefore, be considered interchangeable, especially when at or near the start of a word to be found in a gazetteer.

- Is the name in Latin or has it been Latinized? This practice was common in the past and many historic specimens, especially from Europe, show the name of the country or district in this form. Although such names may seem unfamiliar as place names to the modern reader, many may be recognised as specific epithets: Anglia = England – *anglicus*, Caffraria = S Africa – *caffer*, Gallia = France – *gallicus*. A helpful list of such names can be found in Stearn 1983: 214–234.

- What if the place name is common and has more than one entry in the gazetteer? It will be necessary to work out at least enough of the collector's itinerary in order to select the appropriate entry.

BIBLIOGRAPHIES, BIOGRAPHIES, ITINERARIES AND GAZETTEERS, ARRANGED IN GEOGRAPHICAL ORDER

General

Official Standard Names Gazetteer, United States Board of Geographical Names (country by country in separate volumes).

Campbell, D.C. & Hammond, H.D. (eds.) (1989). Floristic inventory of tropical countries. New York Botanical Garden. Section two: regional reports: 35–455. Covers most tropical (and subtropical) countries, many accounts include notes on topography and vegetation and vegetation maps, and a bibliography.

Desmond, R. (1977). Dictionary of British and Irish Botanists and Horticulturists. Includes all collectors born in the British Isles with brief details of their collecting activities all over the world, and with further references.

Index Herbariorum pt. II (1–7): Collectors. (1954–1988). Regn. Veg. vols. 2, 9, 86, 93, 109, 114, 117. Utrecht. A comprehensive world-wide alphabetical list of plant collectors, together with their geographical areas and dates of activity, and institutions holding their collections.

Stafleu F.A. & Cowan, R.S. (1976–1988)–see **22. Essential taxonomic literature**. This work also includes information on many outstanding botanical explorers and collectors with botanical and travel references.

(**Europe and North America** are not considered in the following list)

East Orient

Wickens, G.E. (1982). A biographical index of plant collectors in the Arabian peninsula (including Socotra). Notes Roy. Bot. Gard. Edinburgh 40: 301–330.

Miller, A.G., Hedge, I.C. & King, R.A. (1982). A botanical bibliography of the Arabian peninsula. Notes Roy. Bot. Gard. Edinburgh 40: 43–61.

Breckle, S-W., Frey, W. & Hedge, I.C. (1969). Botanical bibliography of Afghanistan. Notes Roy. Bot. Gard. Edinburgh 29: 357–371.

— — — (1975). Botanical literature of Afghanistan: Supplement 1. Notes Roy. Bot. Gard. Edinburgh 33: 503–521.

Northern and Eastern Asia

Bretschneider, E. (1898). History of European Botanical Discoveries in China: information on collectors' routes, etc.

Merrill, E.D. & Walker, E.H. (1938). A Bibliography of Eastern Asiatic Botany with supplement by E.H. Walker (1960).

Indian Subcontinent

Biswas, K. (1949). Botanical Survey of India, past, present and future. Proc. bot. Soc. Bengal 13th ann. Sess. 1–23.

Burkill, I.H. (1953–1965). Chapters on the History of Botany in India. 7 parts, Part I being in Journ. Bombay Nat. Hist. Soc. 51, 846 (1953), the remainder in subsequent numbers of this journal. Issued as one volume in 1965. Manager Publications Delhi. This gives information on collectors etc.

Jain, S.K. & Rao, R.R. (1977). Field and herbarium methods. Chapter 1.2 history of botanical exploration in India: 1–6. Today & Tomorrow's Printers and Publishers, New Delhi.

Narayanaswami, V. (1961, 1965). A Bibliography of Indology (Botany) vol. 2. pt. I. A–J, 1961, pt. II, K–Z, 1965. National Library, Calcutta.

Stewart, R.R. (1979). The first plant collectors in Kashmir and Punjab. Taxon 28: 5–12.

— (1982). Missionaries and clergymen as botanists in India and Pakistan. Taxon 31: 57–84.

Watson, M.F. (1992). A botanical gazetteer for Sikkim, Darjeeling and the Chumbi Valley. Royal Botanic Garden, Edinburgh.

Malesia

Fl. Malesiana ser. I, vol. 1 (1950). Cyclopedia of Collectors, with supplements (1958 & 1974): for information on itineraries, localities etc. Later information in Flora Malesiana Bulletin.

Philippines

Tan, B.C. & Rojo, J.P. (1989). The Philippines, A. history of Philippine botany, in Campbell & Hammond: 50–54 (see under **General**).

Sulawesi (Celebes)

Vogel, E.F. (1989). Sulawasi (Celebes), A. Important collectors and herbarium collections, in Campbell & Hammond: 110 (see under **General**).

Papua New Guinea

Frodin, D.G. (1990). Explorers, institutions and outside influences: botany north of Thursday, in Short, P.S. (ed.): 193–215 (see under *Australia*)

Conn, B.J. (1990). Mary Strong Clemens: a botanical collector in New Guinea (1935–1941), in Short, P.S. (ed.): 217–229 (see under *Australia*).

Australasia

Australia

George, A.S, (1981). An annotated bibliography, in Flora of Australia, vol. 1: 14–23. A.C.P.S., Canberra.

Hall, N. (1978). Botanists of the Eucalypts. CSIRO. Supplement (1979), internal report 9; (reprinted 1982); supplement 2 (1989), no. 10.

— (1984). Botanists of Australian Acacias. CSIRO.

Short, P.S. (ed.) (1990). History of systematic botany in Australasia. Proceeding of a symposium held at the University of Melbourne, 25–27 May 1988. Australian Systematic Botany Society Inc.

Willis, J.H. et al. (1986). Australian plants: collectors and illustrators 1780s–1980s. Western Australian Herbarium Research Notes 12: 1–111.

New Zealand

Adams, M. (1990). The botanical collections of John Buchanan FLS, in Short, P.S. (ed.): 231–234 (see under Australia).

Allan, H.H. (1961–1980). A very useful, chronological bibliography, 1769–1958, included in Flora of New Zealand 1, taken up to 1958 in vol. 1: xiii–xxxiv, continued, with earlier additions for 1959–1968 in vol. 2: xv–xxx, and continued further, with earlier additions, for 1969–1976 in vol. 3: xvii–xxxiii.

Pacific

Merrill, E.D. & Walker, E.H. (1947). Index and Bibliography of Pacific Botany in Contrib. U.S. Nat. Herb. 30(1).

Mill, S.W., Gowing, D.P., Herbst, D.R. & Wagner, W.L. (1988). Indexed Bibliography on the Flowering Plants of Hawaii. Bishop Mus., Spec. Publ. 82.

Motteler, L.S. (1986). Pacific Island Names. Bishop Mus., Misc. Publ. 34.

Tropical Africa

General

Distributiones Plantarum Africanum (1969–) (Meise, Belgium).

Fernandes, A. (ed.) (1962). Collectors in Africa–Comptes rendus IV Réunion AETFAT 1962, Lisbon.

Lebrun, J.P. & Stork, A. (1977). Index des cartes de répartition, plantes vasculaires d'Afrique (1935–76).

White, F. (1983). Vegetation of Africa & AETFAT/UNESCO map: includes Madagascar and Mascarenes.

West Africa

Hepper, F.N. (1976). The West African herbaria of Isert and Thonning. Kew.

Hepper, F.N. & Neate, F. (1971). Collectors in West Africa. Regnum Vegetabile No. 74.

Central Africa and Zaire basin

Bamps, P. (1975). Itinéraire et lieux de récolte de Mildbraed lors de sa première expédition en Afrique Centrale (1907–1908). Bull. Jard. Bot. Nat. Brux. 45: 159–179.

— (1982). Fl. d'Afrique Centrale: Répertoire des Lieux de Récolte.

Hallé, N. (1964). Carte des Localités du Gabon. Flore du Gabon 8: 217–228.

— (1968). Notice sur les herbiers Le Testu conservés au Muséum de Paris. Flore du Gabon 14: 11–16.

Letouzey, R. (1965). Les Botanistes au Cameroun. Flore du Cameroun 7.

— (1968). Collectors in Cameroun:– Letouzey, Etude phytogéographique du Cameroun.

Raynal, J. (1968). Itinéraires et lieux de récolte de Georges Le Testu. Flore du Gabon 14: 17–66.

North East Africa

Hulton, P., Hepper, F.N., Friis, I. (1988). Luigi Balugani's drawings of African plants (James Bruce expedition). Yale.

— — — (1991). Luigi Balugani's drawings of African plants. Balkema, Rotterdam.

Gillett, J.B. (1972). W.G. Schimper's botanical collecting localities in Ethiopia. Kew Bulletin 27: 115–128.

Wickens, G.E. (1972). Dr G. Schweinfurth's Journeys in the Sudan. Kew Bulletin 27: 129–146.

East Africa

Polhill, D. (1988). FTEA Index of Collecting Localities.

Verdcourt, B. (1988–1989). Collectors in East Africa–13. F. Stuhlmann (parts 1–3). The Conchologist's Newsletter 106: 113–117; 109: 181–187; 110: 211–219.

Southern tropical Africa

Exell, A.W. (1960). Flora Zambesiaca vol. 1: (1960): 23–37 History of Botanical Collecting in the Flora Zambesiaca area.

— (1968). Collectors in Flora Zambesiaca area. Kirkia 6: 85–104.

Irish, J. (1988). Gazetteer of place names on maps of Botswana. Cimbebasia 10: 107–146.

For Botswana see also Leistner & Morris (1976) below.

South Africa

Gunn, M. & Codd, L.E. (1981). Botanical Exploration of Southern Africa. Balkema, Cape Town.

Leistner, O.A. & Morris, J.W. (1976). Southern African Place Names. Ann. Cape Prov. Mus. 12.

Official Place Names in the Union and S.W. Africa (1951).

Post Offices in the Rep. of S.A. and neighbouring territories (1962).

White (1983) see *General*, above.

Caribbean

Gooding, E.G.B., Loveless, A.R. & Proctor, G.R. (1965). Flora of Barbados, HMSO, London. Useful information in introduction.

Leiva, A. (1989). Cuba, A. Exploration and floristic inventories between the eighteenth and twentieth centuries, in Campbell & Hammond: 327–331. (see under **General**).

Americas

Funk, V.A. & Mori, S.A. (1989). A Bibliography of Plant Collectors in Bolivia. Smithsonian Contributions to Botany.

Lorence, D.H. & Abisaí García Mendoza (1989). Oaxaca, Mexico, A. Important collectors, in Campbell & Hammond: 262 (see under **General**).

Mori, S.A. & Mattos Silva, L.A. (1980). Herbario do Centro de Pesquisas do Cacau em Itabuna, Brasil. Boletim Técnico da CEPLAC 78: 1–31. Information on collections and vegetation of Bahia.

Prance, G.T. (1971). An index of Plant Collectors in Brazilian Amazonia. Acta Amazonica 1, 1: 25–65.

Projeto Radambrasil. (1973–1983). Levantamento de Recursos Naturais. Ministério de Minas e Energia & IBGE. Vols. 1–32. (Vegetation maps and classification of vegetation).

Rzedowski, J. & G.C. de (1989). Transisthmic Mexico, A. Important collectors, in Campbell & Hammond: 274–275 (see under **General**).

UNESCO (1980). Vegetation Map of South America.

Urban, I. (1906). Vitae, itineraque collectorum botanicorum & in Martius, C.F.P. Flora Brasiliensis 1, 1: 1–211. Lives, journeys of collectors in Brazil and biographical notes on collaborators (in Latin).

Verdoorn, F. (ed.) (1945). Plants and Plant Science in Latin America. A wealth of botanical information. Waltham, Mass.

28. CHECK-LISTS

A geographical check-list is a list of plants occurring within a given geographical region, often with brief information on the distribution, ecology, uses etc., and sometimes with keys.

Other kinds of lists can be assembled for particular purposes. These may include lists of type specimens held at an institution or lists of collectors having specimens deposited there.

Check-lists can be compiled in a relatively short time, because they contain less detailed information than works such as Floras. Therefore they are particularly useful when information is needed quickly as, for example, when details of the plants present in a particular area are required for conservation purposes.

CONTENTS OF A CHECK-LIST

A basic geographical check-list consists of a list of plant species names together with information on their distribution in the region being covered. Both **scientific** and **vernacular** (local) **names** should be provided. The former should include **synonyms** which have recently and frequently been used in the region.

Other information *may* be included as follows:

- **Literature references** to the plants listed, especially including relevant local literature.
- **Specimen citations**. These give the source of the information provided.
- **Habitat** or ecological notes. A very brief description of the locality, soil type, and associated vegetation are particularly useful.
- **Altitudinal data**. The minimum and maximum elevations at which the plants grow should be noted.
- **Distribution**. Larger geographical areas will often be subdivided into either political, vegetational or artificial zones and the distribution of plants in each zone given.
- **Keys**. Keys are a valuable aid to identifying the plants and it is worthwhile including them in the check-list.
- **Descriptions of plants**. Detailed descriptions are not normally provided, although there may be short notes for the more difficult taxa. References may be given to descriptions or illustrations published elsewhere.
- **Uses**. Local plant uses, e.g. medicinal, food, are sometimes included. This

may help to identify plants of potential economic value.

ARRANGEMENT OF PLANT NAMES IN A CHECK-LIST

There are two main ways to arrange plant names in a check-list:

- **Alphabetical**. Names are arranged alphabetically by family and then by genus and species under each family. Although this is the simplest method of arranging the names, it is generally of limited use since it is often not apparent which taxa are related. Alphabetical arrangements based strictly on genera, *without* reference to family, should be avoided.

- **Systematic**. Names are arranged such that related families, genera and species are placed near to each other in the list. Thus related taxa can easily be compared, making correct identification easier. It is best to adopt one of the recognized systematic arrangements that are frequently used in herbaria (see **12. The arrangement of herbarium collections**). An index of plant names in alphabetical order can be placed at the end of the check-list to assist the user further.

COMPILATION OF A CHECK-LIST

Materials. The simplest and most economical way to compile a check-list is to use 150 × 100 mm. index cards. Cards can be shuffled into any arrangement required. Each card should have the species name in the top left- or right-hand corner. Other data on the card should also be arranged in a standard format. If a computer is available it may be used to compile the check-list (see **29. Introduction to computers**).

Sources of information. A number of information sources may be consulted and these are summarized below:

- *Literature*. A wide range of botanical literature (e.g. Floras, Monographs, Revisions) has been published over the years, and there is likely to be something of relevance to the region being covered by the check-list. For example if a check-list was being compiled of the grasses of the Serengeti National Park, Tanzania, it would be useful to consult the Flora of Tropical East Africa, Gramineae account, together with Napper, D.M., The Grasses of Tanganyika and Tadros, T.M., Atlas of the Common Grasses of Tanzania. Enquiries made at a botanical library will usually give an idea of the literature available which will be relevant to the check-list.

- *Herbaria*. Herbarium collections usually contain large amounts of unpublished plant data, including distribution, ecology, uses and vernacular names. Such collections are also useful for assessing variation within a species. However, care must be taken in accepting scientific names written on herbarium sheets, and identifications should always be checked.

- *Determination lists.* Determination lists of plants collected in a particular area can provide the basis for a check-list. Many herbaria retain and file copies of these lists. An index to the collectors and a knowledge of their itineraries is vital.

- *Field work.* This is often a good way of obtaining data, allowing gaps in distributions to be filled, new taxa added etc. Experienced botanists frequently undercollect very common taxa, and these should therefore not be overlooked.

REFERENCES

Steenis, van (1972)

29. INTRODUCTION TO COMPUTERS

Note: Technical terms printed in bold type which are not explained in the text can be found in the **Glossary** on pp. 188–191.

Computers have become increasingly important for performing certain routine tasks in the herbarium. Although computers are, at first sight, complex machines they are in fact quite simple to operate, given some basic training. Basically all that is required is knowledge of the computer keyboard, how to use specific programs, and how to operate the printer.

USES, ADVANTAGES AND DISADVANTAGES

Some of the main *uses* of computers in the herbarium are:

- Preparation of labels (see **6. Label design and production**).
- Preparation of loan forms and management of loans.
- Preparation of determination lists, which can be continually updated when required.
- Recording incoming material.
- Cataloguing herbarium specimens.
- Preparation of correspondence, especially variants of a basic letter and production of address labels.
- Production of bibliographies.
- Production of check-lists.
- Specimen-based **databases** can provide a rich source of information for researchers such as ecologists and conservationists, thus saving staff time when answering enquiries in these fields (see below, *Database management systems*).
- Botanical research, e.g. flora writing, identification keys, numerical systematics, phenetic and cladistic analyses, and distribution mapping: these important applications are beyond the scope of this book and are therefore not covered here.

Apart from the above uses to which computers can be put, the following *advantages* should also be considered:

- They store large amounts of information which, provided care is taken over how the data is stored, may be easily and flexibly available (retrievable).

- They can allow the simplification of routine herbarium recording functions so that staff time can be more profitably spent on other work.

- They may help to achieve consistency of format and content.

- They make the direct communication of data with other organisations more rapid, *providing* the equipment and software are compatible.

Although computers can ease the burden of certain jobs, before deciding to use a computer for any herbarium task it is worth remembering that there can also be *disadvantages* under certain circumstances. Some problems which may be encountered are:

- Initial cost and subsequent maintenance. Although personal computers have been reduced in price over recent years, a complete system with a printer and software could cost more than herbaria with limited budgets can afford. Also, periodic maintenance and repairs need to be carried out by specialist engineers, and in some countries specialists may not be easily available.

- The labour and amount of time which can be taken up in putting data into the computer. The use of manpower on such a task may not be justified, particularly in herbaria with few staff.

- Poorly designed systems will result in data being stored in cumbersome and possibly unuseable forms.

- System and human errors: errors in either the computer program or the data may easily occur, and once in the system they may be difficult to find and remove.

- The cost of support expertise to help staff out of difficulties with general computer usage or program failure.

- Most standard software packages require considerable and tedious adaptation to specialist activities. This generally requires programming which is skilled and time-consuming.

- Although the general standard is improving, many manuals (instruction books) supplied by manufacturers of both hardware and especially of software have in the past been poorly written, difficult to understand or even misleading. For a beginner, remote from support help, this problem can be immense. The extra expense of purchasing a useable manual may be necessary.

- There can be a significant cost in training before a person is competent to use a computer and software.

PURCHASING A SYSTEM

A computer system requires both **hardware**, the actual computer and ancillary equipment (or **peripherals**), and **software**, the **programs** that are used to run it. In order to assemble a fully satisfactory system it may well be necessary to spend as much on the software as on the hardware. Because computer systems are constantly becoming cheaper and more effective, it is always difficult to decide whether or not to delay purchase. Is postponement of the work worth a possible cost-saving?

Before purchasing a computer system for the first time it is vital:

- First to define your requirements. What is the system to be used for and will it save staff time (i.e. be cost-effective)?

- Then to decide what software will be needed.

- Finally to decide what hardware will be needed.

Aim for a system that is neither too simple nor too advanced for the necessary work; some computers can be enhanced at a later stage if necessary. Check that all items of hardware and software function together. It is important to seek advice from a specialist or dealer; new and improved designs are constantly entering the market and it is important to have the most up-to-date information. Find out from the dealer if discounts are possible and what services are available. If there are no staff with computer skills in the herbarium, ensure that the dealer is prepared to supply the system already **configured** and with the software **installed**. These tasks can be very difficult for a computer novice to perform and may not automatically be carried out by the dealer without an agreement. Ask what help facilities and after-sale services are available. Most software companies provide telephone 'hotline' support for their programs. In choosing a system the level and costs of services may be as important a consideration as the choice of the system itself. Mail order firms can be relatively inexpensive and often offer a good level of after-service. If all other considerations are equal, choose a machine made from standard industrial components in preference to one made from components exclusive to one manufacturer, which would be more expensive.

The following basic notes on *hardware, computer consumables* and *software* should give some idea of what could be available, and may raise points to consider before the choice is made.

HARDWARE

Computers

Computers are of various sizes. The largest are **mainframe** computers, which are very powerful machines housed in special rooms with numerous **terminals** available (even thousands) in the staff working areas. They are very expensive and require frequent maintenance by specialist personnel. Medium sized computers, known as **minicomputers** (or **minis**) can be described as small scale mainframes; they can support hundreds of terminals. Mainframe and minicomputers are appropriate for larger institutions and these will not be further discussed here, however, see *Network* below. The smallest computers are known as **personal computers** (or **PCs**; also **microcomputers** or **micros**). PCs are available in many different sizes and designs (see below), but there are two main types – the IBM (or IBM compatible) and the Apple Macintosh, which differ in the kind of processor used and the software available. If it is necessary to transfer software or peripheral equipment from one computer to another they must be compatible and files must share a common data format. Compatibility has presented many problems in the past but it is now possible, by various means, to overcome most of these problems at a technical level. Apple Macintosh supply software which converts IBM data files for use with Macintosh machines as standard with all their computers.

It is important to consider the storage memory of the computer. Memory is measured in **bytes** (see **Glossary,** also *bits, KB* and *MB).* As a rough estimate, a 1000-word document would require about 6000 bytes (or 6KB) to store (or 12KB if a back-up is made). The capacity of **hard disks** (see below) is measured in megabytes (MB). The power (capability and speed of operation) of computers varies according to the processor (chip) and the RAM (see **Glossary**). Choose a computer with RAM of at least 1MB.

The two basic models of PCs are:

- *Desk-top*: mains powered computers.
- *Portable* (*lap-top*) computers: mains and/or battery powered. Compact versions are called *notebook* computers and even smaller models, *handbook* computers.

At present portables can compare favourably in capacity and reliability with most desk-top machines but they are more expensive than the equivalent desk-top models.

Network

PCs can be linked together to form *networks.* These may vary in size and complexity from two machines in the same room to hundreds of machines distributed throughout an organisation. A network requires special operating system software (see p. 183) (of which 'Novell' is probably in widest use) and would normally have one or more, much larger, more powerful PC's (**servers**) providing central facilities for data storage and application programs (see p. 184).

Networks have many advantages over collections of stand-alone PCs:

- Expensive items of hardware (laser printers, data storage, back-up facilities) can be shared by many users – even those with cheap machines.
- Common software licences may be cheaper than multiple single-user licences.
- Communication and data exchange among users is easier.
- Data back-up and system security can be centralized.

On the other hand networks can have some drawbacks:

- Additional skills and experience will be needed by some staff who should be given responsibility and allowed the time to provide central support and network management.
- Network licences for software, if only intended for one user, would be more expensive.
- The development of databases to run over a network as multi-user applications will add to the complexity of the development of the necessary software.

The following items need to be considered as part of a standard PC system:

Disk drives

Information is normally stored in magnetic form on disks. The machinery necessary for holding and operating the disks is the *disk drive,* which is often incorporated in the computer but may also be a separate unit. The two kinds of disk most commonly used are:

- *Diskettes* (or floppy disks; see also below, *Computer consumables*). These are inserted into the disk drive. Currently there are two sizes, 5¼ in. and 3½ in. in common use, but 3½ in. is becoming the standard; two sizes, 8 in. and 3 in. are obsolete. The different sizes must be used in the correct-sized disk drives; they are not interchangeable.
- *Hard disks* (sometimes known as Winchester disks) are rigid disks that are fixed inside the PC. They contain much more storage than diskettes, with a

range of capacities, at least 40 MB is recommended. Hard disks operate at a faster speed than diskettes.

For flexibility in use, a computer fitted with a hard disk and disk drives for both 5¼ in. and 3½ in. diskettes is best; failing this, a hard disk and 3½ in. disk drive should be chosen. Drives for **tape back-up** may be fitted to higher capacity machines or may be purchased as separate units. **Bernoulli disk drives**, available as separate units, may also be used for **backing-up** (see below).

- *CD-ROM reader* (or disk drive). An attachment for reading **CD-ROM** can be purchased for use with PCs. This will become increasingly important as reference literature, such as the Index Kewensis, is marketed in this form.

Monitors or **VDU**s (visual display units)

These look similar to television sets. The standard monitor at present is the **VGA** (video graphics array) with a high resolution and up to 256 colours, although they may also be black and white. The **EGA** (enhanced graphics adapter) is becoming obsolete, while **CGA** (colour graphics adapter) and monochrome monitors are virtually obsolete already, but are still a viable cheap option in some countries.

Portable computers have built-in monitors, so a separate unit is not needed. The cheaper units are usually of the monochrome liquid crystal display (**LCD**) type but other kinds are available. Colour models based on other technologies are becoming available but are mains rather than battery powered and still expensive, although becoming cheaper.

Printers

These are needed to provide a copy of the data on paper. Printed copies of data held in a computer system are referred to as **hard copy**. The speed of printing is measured in **cps** (characters per second) or **ppm** (pages per minute) depending on the type of printer. The quality of printing also varies. Some kinds of printer are capable of colour printing but this is unnecessary for most herbarium applications and probably not worth the extra cost (currently considerable).

Daisy-wheel printers. The printing heads, which are interchangeable, are wheel-shaped with the characters at the ends of spokes. The character range is limited and the printers are rather slow – 20–60 cps. The print quality is generally high but limited by the type of ribbon. Daisy-wheel printers are usually noisy and are rapidly becoming obsolete.

Dot matrix. Printers operating on the dot matrix principle print characters made up of variable patterns of electronically controlled dots within a rectangular matrix (dot matrix). The character set is therefore not limited, and a choice of fonts (or type face), pitch (horizontal spacing) and graphics is possible. The quality of print is governed by the number of dots used to create the character. The following types are available.

- *Dot matrix printers* (or *pin writers*). The print head contains a matrix of pins controlled by electromagnets which strike the paper, thus building up the characters. The two main types have either 9 or 24 pins in the print head. The ink is supplied by a ribbon. Speeds vary from 40 to 150 cps, or even c. 300 cps for some very high speed 9-pin printers. They are usually noisy, although some can operate more quietly at a slower speed.

- *Ink-jet printers.* In this type of printer the dots are formed by sending a stream of tiny dots of electrically charged ink onto the paper. The printer is very quiet as there are no pins to strike the paper. Monochrome printers can reach speeds of c. 500 cps. They give a higher print quality than dot matrix printers and are intermediate in price between dot matrix and laser printers.

- *Laser printers.* Because of their non-impact action laser printers are quiet. Being page printers they can usually produce 6 or even up to 15 pages per minute. They have a high resolution, usually 300–400 **dpi** (dots per inch) and therefore produce high quality print. With the use of font cartridges they can produce a wide range of sizes and styles of type, symbols and graphics but many printers now have numerous **scaleable fonts** installed by **default**. When used with a *Desk Top Publishing (DTP)* printer control system they can offer far greater flexibility in style of printing than dot matrix or ink-jet printers.

Keyboard

The user usually communicates with the computer via a keyboard which resembles that of a typewriter. In addition to the usual characters present on a typewriter a computer keyboard will have various function keys and often a number pad.

Optional Peripherals

Acoustic Hood. A sound-proofing hood to place over daisy-wheel or dot matrix pin printers which are too noisy for the intended place of use.

Paper feeders to accommodate separate sheets of paper rather than continuous computer paper.

Mouse. This is a small device which is moved over a flat surface to control the position of the **cursor** on the screen. The mouse is provided with controls for the selection of functions. It is mostly used with **windows** or with graphical programs.

Scanner. An input device that allows images (pages of text or graphics) to be read and stored by a computer. An **image scanner** converts documents, drawings or photographs into a digitised machine-readable form. Many scanners have an

optical character reader (**OCR**) facility which scans printed or written characters and converts text into characters that can be recognised by a PC. Scanned input, however, uses a lot of storage space. There are three basic designs of scanner: flat-bed (with a stationary target and moving optics), sheet-feed (with stationary optics and a moving target), and camera-scanners (used for three-dimensional objects and large flat images).

Modem. A device that allows data to be communicated to another computer via the telephone system, or for connecting terminals to a mainframe or mini-computer.

Uninterruptable power supply (**UPS**). If the electricity supply is liable to interruption or changes in voltage an UPS will be an essential purchase. This apparatus will filter the mains power supply and gain the user time to **power down** the PC if the supply fails completely, as it contains batteries. The battery life is between 20 minutes and 4 hours.

Maintenance contracts. These should be considered for all items of hardware. Usually they can be arranged through the supplier or manufacturer, but they can be costly, about 12–15% of the cost of the hardware.

COMPUTER CONSUMABLES

Diskettes

A supply of 3½ in. and/or 5¼ in. diskettes will be needed for back-up purposes. Check which density of diskette is required. Both sizes are available in Double or High Density. Some machines can accept both densities.

5¼ in. diskettes are cardboard-encased and have a protective paper sleeve; the 3½ in. design is more robust. Storage boxes for diskettes will be found useful. New diskettes must be **formatted** before use.

Special cassette **tapes** may be needed if the computer has a tape back-up facility.

Stationery

Computer paper is generally of high quality (see **5. Materials**) but some is re-cycled and this should only be used for draft purposes. Two main formats are available:

- Continuous (or sprocket-fed) paper, for dot matrix pin printers, is perforated at sheet-length intervals and sometimes along the edges to allow the removal of the punctured strips which allow traction with the printer.

- Single sheets of paper can be used in all types of printer, and must be fed individually if the printer is not fitted with a sheet-feeder. High grades of copy paper should be chosen for use with laser printers.

Sheets of self-adhesive and gummed labels are available for most types of printer, but for laser printers particular specifications are required for such items because of the high temperatures and pressures involved. If the labels are intended for long-term herbarium use, archival quality is necessary. Ordinary self-adhesive labels are only suitable for temporary office use, e.g. addressing envelopes. Most printers are capable of taking a range of weights of paper, including cartridge paper and thin card suitable for cabinet labels and index cards etc. However, trials with the printer may be necessary.

Inks. The ribbons in dot matrix printers, the ink cartridges for ink-jet printers and the toner cartridges for laser printers will need replacing. It is good practice to keep a moderate supply, especially if they must be ordered from overseas. Manufacturers' warnings over re-use or refilling containers should be followed. Inks for computers are generally *not* of proven archival quality and some have been known to dark fade (i.e. fade even when not exposed to light).

SOFTWARE

Computer programs are known as software. The software instructs the hardware how to perform. It must first be **installed** in the computer, i.e. loaded and configured to the system's capabilities.

Operating systems (OS)

The operating system runs the computer for the user; it controls the basic low-level hardware operations and **file** management. The operating system is usually supplied with the computer as a software package (often installed).

The kind of operating system supplied with IBM or IBM compatible computers is known as **DOS** (Disk Operating System). There are different versions of DOS; 3.3 is still widely used but 5.0 is the most recent and best. It is important to know which version of DOS is supplied with the computer, because recent versions of applications programs (see below) may not be compatible with early versions of DOS.

Some knowledge of how to control the computer by **commands** may be needed, e.g. it is especially important to know how to format and copy diskettes. Before the computer is used for the first time it may be necessary to execute **utility programs**, which control functions such as file searching, file copying etc., and **device-driver programs**, which are used to interface and manage an input/output (I/O) device or peripherals, e.g. a mouse. The manufacturer's manual will give instructions but if necessary get help from the supplier (see above, *Purchasing a system*).

In Apple Macintosh computers all operations are carried out via a user-friendly **menu** system or **windows**. Software packages (e.g. 'Windows 3') are now available for IBM computers which permit operations to be carried out in a similar way.

Applications programs

Programs which are designed to allow a particular task to be carried out are known as **applications programs**. Word-processing programs, database systems, office packages and even computer game packages are all examples. Applications programs may be available as commercial packages or they may have to be specially tailored (see **Glossary**, *Programming languages*).

Software will constantly improve and the most recent versions will have additional functions; however, they can often only be used with the more advanced versions of the operating system on high specification PCs. Recent versions will normally allow conversion of data from an earlier version, but this can be a complicated procedure. There are also versions for IBM compatible and Apple Macintosh computers.

Software (apart from the operating system) must usually be purchased. Commercial products can be very expensive, but discounts are often available for educational or research institutions. Commercial software is copyright but 'public domain' programs have no copyright. Privately written programs of limited distribution are also available, usually for sale. Pirated (i.e. unauthorized) copies must be avoided; in most countries pirating is illegal (see below, *Data and system security*).

The two most important commercially available applications programs for herbarium management (*excluding* taxonomic botanical research) are word-processing and database management systems. Some specially designed programs may combine aspects from both.

Word-processing (WP)

A word-processing program (such as '*WordStar*' or '*Word Perfect*') allows a computer to be used like an electric typewriter but it is far more powerful. The program makes it possible to mark a block of text, then move, copy or delete it as required, or to insert a new portion of text without having to retype the whole. There may also be facilities for checking spelling, searching for selected words and changing them if required, or for arranging simple lists in alphabetical order. Modern WP packages allow graphics to be combined with text and include some desk-top publishing facilities.

Many herbarium tasks which require several versions of a basic text (e.g. correspondence requesting loan returns or data labels) or the constant updating of simple lists can be efficiently carried out with a word-processor.

If files compiled on different word-processing packages need to be combined there are programs enabling this to be done. Alternatively, a scanner can be used (see above, *Scanners*) to convert the printed file back into **ASCII** text.

A *desk-top publishing* program can take word-processed text and assist in formatting it into pages in a defined way, and allow graphic images and text to be combined. A range of fonts may be available and, when combined with a laser printer, pages suitable for publication can be produced.

Database management systems (DBMS)

Pankhurst (1991, p. 11) defines a database as 'data which has structure together with some system for organising it'.

A database management system allows items of data concerning a subject to be stored in different **fields** and for these fields to be accessed and organised in a variety of ways. As a simple example, the data shown in fig. 17 (p. 63) could be **captured** (i.e. copied) as follows:

Field	*Data for A*	*Data for B*
Family	Rubiaceae	Rubiaceae
Genus	Mussaenda	Gardenia
Species	erythrophylla	augusta
Country	Uganda	Kenya
Collector	Other, A.N.	Other, A.N.
Number	123	456
Year	1960	1985
Month	5	4
Date	15	1

With a DBMS the data could be organised to obtain, for example, the following kinds of information:

- A numerically arranged list of all specimens collected by A.N.Other with their determinations.

- A list of all specimens of Rubiaceae (arranged alphabetically by genus) from Uganda and Kenya.

- A list of all *Mussaenda* specimens collected in Uganda.

Modern DBMS are very powerful and can work with many files simultaneously; they can interface with other kinds of software such as word-processing systems and spread sheets (tabulated formats). At present the most widely used commercial system for herbarium activities is 'dBASE', although it is not the most powerful.

Once a database has been set up and the fields and number of characters within each field established they may not require a high level of skill to operate, but this depends on the elegance of the system. However, the design of a database requires more knowledge, and may prove difficult and time-consuming for a beginner. For this reason many privately produced customized systems have been developed, usually using 'dBASE' (or other commercial systems) as a database language.

Customized systems have been developed to help perform a range of documentation tasks in the herbarium, facilitating label production, collection management and recording transactions. These activities are usually performed with **specimen-based** systems; other types of system (e.g. taxon-based or name-based) are used in herbaria but will not be considered here (see Beaman & Regalado 1989; Abbott et al. 1985). More powerful specimen-based systems allow data to be extracted for purposes such as taxonomic research and flora writing. As yet there is no universally accepted way of standardizing databases and no universally adopted system for herbarium use. It must also be remembered that standardization of hardware or database software is irrelevant if the data terminology are not standardized, e.g. whether British Isles, England, Great Britain or U.K. is used.

The following points should be borne in mind:

- Database projects are usually long-term and will require long-term funding.

- Can the data-set be maintained independently of future developments (including redundancy) of the software or hardware?

- The choice of data should be given as much thought as the software. Only high quality data should be stored; data that is inconsistent or may become redundant can cause difficulties in databases (see Allkin & Maldonado 1991).

- Does the software ensure a sufficiently high consistency of stored data? For example. 'DePew' may not be retrieved if it were erroneously entered previously as 'De Pew'.

- Is the system required solely for use within the institution, or should it be possible to export the data stored in it to systems in other institutions?

- How sophisticated should the system be? Try to avoid one-off personal solutions or complicated systems that can only be operated by one member of staff who could leave.

- Find out from colleagues and visitors from other institutions what systems are available, how well they perform and how easy they are to obtain and use.

- Try to be aware of the latest developments by consulting literature, magazines and newsletters (e.g. Museum Documentation Association (MDA Newsletter) & American Society for Systematic Collections (ASSC Newsletter)). Reviews of customized systems are often published in botanical periodicals (especially Taxon).

Unauthorized use of computers, programs and data must be prevented.

- Mainframe computers usually have pass-word systems or keys (physical or electronic) to prevent use by unauthorized personnel, and such devices may also be present on PC computers.

- Many commercial manufacturers of software protect their programs with anti-pirating devices (or 'copy-protection schemes') so that the data cannot be illegally passed or sold to other would-be users. Some copy-protection schemes create hidden files which can cause problems with the hard disk especially if used with various utility programs. Because of campaigns against this practice, most mass market packages are no longer copy-protected but many specialist packages still are. If software is copy-protected a back-up copy should be kept.

- People with malicious intent have developed a series of programs, known as **viruses** which cause havoc to the software and often lead to a destruction of the data. They are obtained either from using illegally marketed software (especially computer games) or by copying files from an infected system. They are passed from computer to computer in a way analogous to a virus amongst people. Only genuine software packages should be used and any files to be copied onto a computer must be run against a **virus check** program (on a separate system) *before* the data are accepted. Utility programs exist that permit the detection and elimination of viruses from PCs.

It is essential that *all visitors* bringing their own diskettes into the herbarium are asked to submit them for virus checking before they are put into a computer belonging to the institution. Even if visitors have brought their own portable computers they may still wish to use their diskettes in the institution's hardware in order to print.

Back-up

A copy (or copies) of a file or data or disk should be made on a diskette, tape or Bernoulli disk in case the original is damaged, corrupted or lost. These are known as back-up copies.

Individual data files should *frequently* be backed-up with the current version. This is especially important if the electricity supply is subject to changes in voltage or interruptions; data can be deleted instantly in these circumstances (see above, **Uninterruptable power supply**). A daily back-up should be made; the use of a separate diskette designated for each working day of the week is recommended. In addition it is important to make back-up copies at monthly or six monthly intervals. It may also be necessary to keep back-ups of the entire hard disk; this may be done on tape or Bernoulli disk if the facility is present.

Back-up copies should be stored in a different location from that of the computer. Daily copies can be kept in a different room in the same building, but monthly or six monthly copies should be stored in a different building, preferably at a different locality. Diskettes should be stored away from heat sources, high humidity and strong magnetic fields (e.g. near telephones or loudspeakers). They are, however, not affected by airport X-ray machines. See DePew (1991): 230–231.

Housekeeping

Tasks that have to be regularly carried out to maintain a computer system are referred to as housekeeping. The most important housekeeping activities are:

- *Deleting unwanted files.* Files that are no longer required should be deleted from the **directory** so that the storage space is made available for new material. Care should be taken not to delete the wrong files; this can happen if files have been given similar names by different users.

- *Maintaining file directories.* Files can more easily be lost or accidentally deleted when they occur in directories together with many other files. Orderly and logical groupings of files should be kept in separate subdirectories, which can be nested within one another. In general, no more than 40–50 files should be stored in one subdirectory.

GLOSSARY

This includes both words used in the text and additional computer terms in common use.

ASCII (American Standard Code for Information Interchange). A standard method of representing alphanumeric characters (i.e. letters or numbers) in binary code. An **ASCII character** is a character which occurs in the list of ASCII codes, i.e. any letter (upper or lower case), number, punctuation mark etc., usually present on the keyboard. The whole set contains 96 printed characters. Additional characters (e.g. é, ç, ø, etc.) may be present in the **extended character set**.

Bit (binary digit). The smallest unit of data that a system can handle. It can exist in only two states, '0' or '1'.

Byte. A group of 8 **bits** which a computer operates on as a single unit; it gives the equivalent memory storage of one character. See also **KB** and **MB**.

CD-ROM (Compact Disk – Read Only Memory). A commercially produced compact disk storing large amounts of data, e.g. bibliographies, dictionaries etc. The disk can be read but data cannot be added to it. The data is stored optically rather than magnetically. A separate CD-ROM reader or disk-drive is required for PCs.

CGA (Colour Graphics Adapter). See *Monitors* in text, p.180.

Command. A word or phrase which is recognized by the computer system and will start or stop an action.

Configure. To set up the system (hardware and software) to particular specifications or requirements by commands.

CPS (characters per second). A measure of the speed of a printer.

CPU (central processing unit). The basic computer unit, i.e. without peripherals.

Cursor. A marker on the display screen which shows where the next character will appear. Usually it will be a bright flashing square or underline.

Database. A collection of data relating to a subject. See text, p.185.

DBMS (Database Management Systems). See text, p.185.

Default settings. The basic settings for software, printers etc. on leaving the factory.

Device-driver. Program used to interface and manage an I/O device (see below) or peripheral, e.g. mouse.

Directory. A list of the names of files stored in the computer; it may also give other information such as date and time and amount of storage space left on hard disk.

DOS (Disk Operating System). See *Operating systems* in text, p.183.

EGA (Enhanced Graphics Adaptor). See *Monitors* in text, p.180.

Field. An area (e.g. a column) under a heading into which a particular type of data is put in a database, e.g. collector's number, date or family.

File. A single unit of related records, e.g. list of addresses, collectors etc.

Floppy disks. See '*Diskettes*' in text, p. 182.

Font. A set of characters in the same typeface and size.

Format (as a DOS command). To make a blank diskette ready to receive data, by writing control and track location information on it. This is done by inserting the diskette into the disk drive and giving the appropriate simple command.

Hardcopy. Printed copies of data.

IBM (International Business Machines Corp.).

Install. To set up a new computer system to the user's requirements or to configure a new program to the existing system capabilities.

I/O (input/output). The receiving or transmission of data.

I/O device. A peripheral such as a terminal.

Kb (kilobit). 1024 bits.

KB (kilobyte). 1024 bytes.

LCD (Liquid Crystal Display). See monitors in text, p.180.

Load. To transfer a file or program from a hard disk or diskette (or tape) to the main memory.

Mb (megabit) = 1,048,576 bits.

MB (megabyte) = 1,048,576 bytes.

Memory. The long-term storage capacity on a computer system (on the hard disk plus diskettes etc.); also short-term storage (or main memory), see RAM.

Menu. A list of available options displayed on the screen.

MS-DOS. The version of DOS produced by the Microsoft company.

OS (operating system). See text, p.183.

PC-DOS. The version of DOS supplied by IBM.

Pixel. The smallest unit or point of an image on a monitor screen.

Power down. Including: exiting from the program (if this is not done the data will be lost); logging off, if necessary; and finally switching off the power.

Program. A set of instructions which directs the computer to carry out a particular task.

Programming languages. Permit the writing of instructions in the form of specially designed applications programs. The most commonly used language for PCs is '**BASIC**' but this is being superceded by '**C**'. Other programming languages include '*Pascal*', '*Logo*' and '*Cobol*'. Programming is a task for the experienced computer user, not the beginner.

RAM (Random Access Memory). Sometimes called read/write memory. New data is entered into the RAM; if not saved, it will be lost when the power is switched off. See also **ROM**.

Resolution. The number of pixels a monitor screen or printer can display per unit area – measured in dpi (dots per inch). The higher the dpi the better the image quality.

ROM (read only memory). This kind of memory was written into the computer at the time of manufacture. The computer can read it but no additions are possible; when the power is switched off it will not be lost. See also **RAM**.

Scaleable font. On a laser printer, a given font that can be produced in a range of sizes.

Softcopy. Text displayed on monitor screen.

Storage. The amount of space available for storage of data. This can be space on diskettes as well as space inside the computer. See also **Memory**.

Terminal. A device, usually consisting of a monitor and a keyboard, which allows the entry and display of information to a central computer system (mainframe, mini or network). A **dumb terminal** has no processing facilities. An **intelligent** (or **smart**) **terminal** contains a CPU and memory.

Trojan horse. A program that appears like a normal applications program but which conceals an element, which when triggered, can destroy the system. See below, **Virus**.

Utility. A program that is concerned with routine activities such as file searching, copying files, file directories, etc.

VDU (Visual Display Unit). See *Monitors* in text, p. 180.

VGA (Video Graphics Array). See *Monitors* in text, p. 180.

Virus. A hidden program with a routine which corrupts all data and files. It spreads from computer to computer when diskettes are exchanged, see p. 187.

Window. A section of the screen used to display information and give options that can be selected. A window can be superimposed on the information already on the screen whenever needed.

Worm. A self-replicating program which can slow the hard disk and eventually fill it; it is transmitted from computer to computer like a virus (see above).

WORM (Write Once Read Many times memory). A WORM-drive enables data to be written on a blank optical disk – see CD-ROM.

REFERENCES

Abbott et al. (1985)
Allkin & Bisby (1984)
— & Maldonado (1991)
Beaman & Regalado (1989)
DePew: 230–231 (1991)
Jury (1991)
Macrander & Haynes (1990)
Morin et al.: 135–141 (1989)
Morris & Manders (1981)
Pankhurst (1983)
— (1991)
Reznicek & Estabrook (1986)
Roberts (1985)
Sarasen (1981)
— & Neuner (1983)
Wetmore (1979)
Williams (1987)

30. COLLECTING AND PRESERVING SPECIMENS

All collecting expeditions, especially those to another country or remote area, must be carefully and thoroughly planned. It is vital that all permits granting permission to collect and export specimens are obtained. The logistical details cannot be covered here, but see: Fosberg & Sachet 1965: 45–50; Jain & Rao 1977: 22–33; Matthew 1981: 73–81; Womersley 1981: 123–16.

The importance of collecting good herbarium specimens cannot be overemphasized. A small number of really well preserved and annotated specimens is far more valuable than a large number of poor specimens. Each collection should fulfill the following requirements:

- It must be as representative as possible of the population or show as wide a range of variation as possible and be accompanied by an unambiguous collection number (see below, *Collecting and numbering specimens*).
- It must have good collecting data recorded (at the very least: collector with number, date, locality, habitat and any plant details likely to be lost on drying) (see below, *Recording data*).
- It must be well preserved (see below, *Processing specimens*).

COLLECTING AND NUMBERING SPECIMENS

Equipment required for collecting and numbering specimens

- *Tools etc.*, pocket-knife, long-bladed knife (machete, parang, facão or panga), secateurs, scissors, long-handled pruner, pruning saw, axe (or bow saw), spade (or entrenching tool), trowel, chisel and hammer, thorn-proof gloves.
- *Polythene bags* (assorted sizes and thicknesses).
- *Field portfolio*, i.e. two lightweight boards (slightly larger than a herbarium sheet and preferably hinged along one length) containing paper folders and held together with straps. Used for storing plants that wilt too quickly to be carried in a polythene bag (see plate 1, B).
- *Number tags* (jeweller's tags) and pencils – used for adding the collection number to each specimen.
- *Plastic bottles* (assorted sizes with wide mouths).
- *Liquid fixative* or *preservative* (see **31. Collecting material for ancillary disciplines**) – for collecting delicate material or separate organs.
- *Binoculars* for observing forest canopy.

Ideally, as each specimen is collected it should either be 'field-dried' or placed in a folder and processed according to the 'alcohol method' (see below, *Processing collections*). In practice there is seldom sufficient time in the field to permit this but collecting can be speeded up by placing the specimens in polythene bags, and pressing later; the shorter the delay, the better. Use separate bags for each habitat or locality, and annotate each clearly as to its contents. Small and fragile plants should be treated with care; place them in a small bag within the main bag. Specimens of the same or similar species should also be separated to avoid later confusion. Not all taxa (e.g. *Impatiens*) are suitable for this treatment and a field portfolio should be used in such cases.

What and how much to collect

First look around and become familar with the habitat; a better idea of frequency and availability of species will be gained. Material truly representative of the population can then be collected.

Decide which parts should be collected, and how the morphology, size range, and other features of the whole plant can best be represented and recorded in the herbarium. A good collection should comprise adequate samples of all the organs available and all stages of development. It is important to select the material so as to retain as much information as possible. The following points should be borne in mind:

- The supporting branch should be cut so that the attached petioles, axillary buds and any stipules are retained (see fig. 46). Always try to keep compound leaves intact.

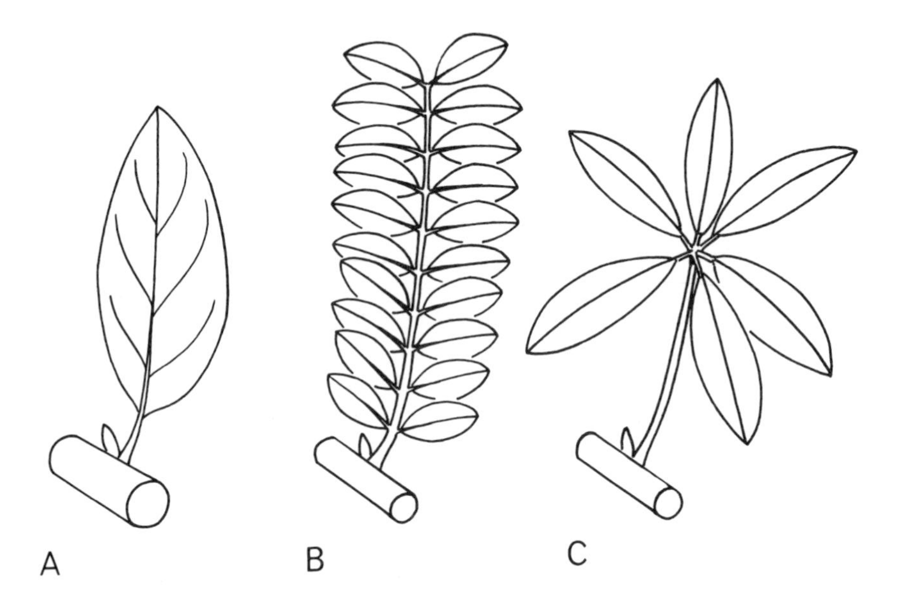

FIG. 46 — Cut to keep petiole attached to stem. **A** simple leaf; **B** pinnate leaf; **C** palmate leaf.

– Cut woody subjects so as to demonstrate as much of the branching pattern as possible. Twiglets detached from the supporting stems or the branch alone cannot supply this information. Where possible the stem apex should also be retained (see fig. 47).

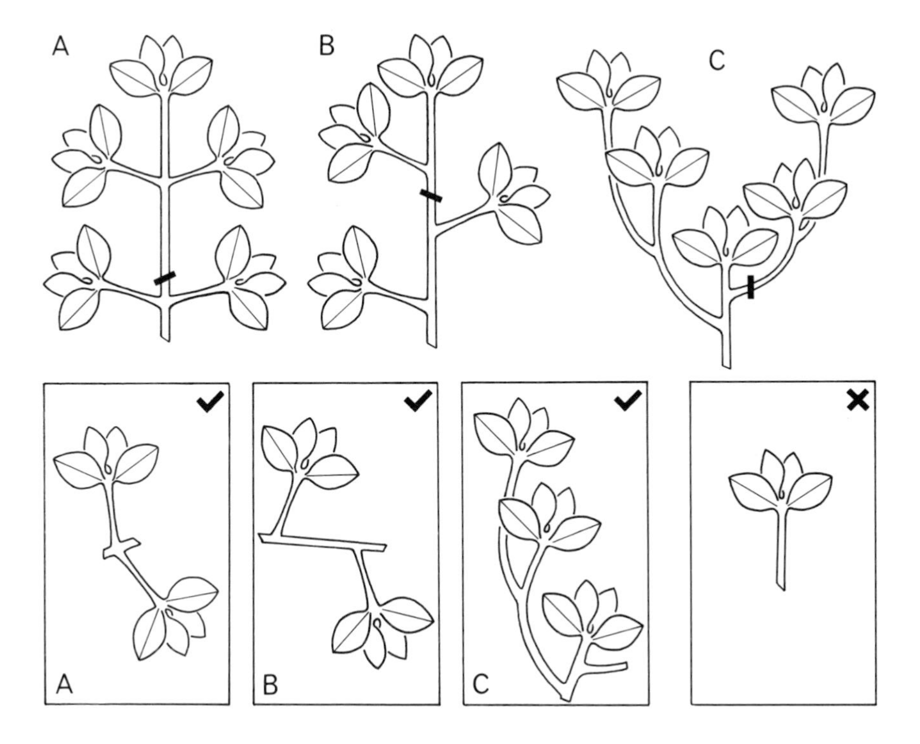

FIG. 47 — Cut to keep as much of the branching pattern as possible. **A** opposite; **B** alternate; **C** sympodial. The incorrect example could have come from any of these.

– Look out for:
  – *heterophylly* (different shaped leaves), including juvenile foliage or shade leaves. State which kind.

  – *monoecious* plants (♂ and ♀ flowers on the same plants). Try to collect both flower types.

  – *dioecious* plants (♂ and ♀ flowers on separate plants). Collect each sex under a separate number, or if a population of small plants, collect sufficient of both sexes for all duplicates.

  – *heterostylous* plants (⚥ flowers with long-styled and short-styled forms). Collect each form under a different collecting number, or if a population of small plants, make sure there is sufficient of both kinds for all the duplicates.

- Take special care that each collection number consists of only one taxon. If there is doubt, this must be noted and, if possible, separate collections made.

- Portions cut off a large plant should be big enough to fill a herbarium sheet. Often it is possible to fit an adequate sample on one sheet, but do not be tempted to collect only the smallest leaves etc. just because they fit the sheet.

- If the plant is large, collect specimens (with field data) complete enough to be fully useful. These may fill several sheets (see below, *Large* and **32. The collection of pteridophytes**).

- If the plants are small, collect sufficient individuals to fill the sheet.

Wherever possible additional loose flowers and fruits should be included and placed in packets if appropriate (see p. 216, *Arranging and preparing the specimens*). If seeds are abundant and fully mature, they may be dried and a packet-full placed with the specimen. Write the collecting number on the packet.

Duplicate sets of all specimens should be collected whenever possible. The number of duplicates aimed at may depend on agreements with other herbaria, especially the host country, drying capacity etc. (see p. 224, *Selection of duplicate sets* and **18. Duplicate distribution**). For a relatively undercollected area, 5–6 sets would a reasonable number.

If any species is known to be rare, conservation must be kept in mind. A small sample accompanied by photographs or photographs alone may be adequate (see **36. The collection of living material** and **41. Conservation and the Herbarium**).

Some types of plant require special treatment; the nature of some organs also dictates careful handling. For specimens that are:

- **Succulent or fleshy**. Cut longitudinally and/or transversely; sometimes it may be necessary to scoop out the inner tissue. Consider if half of the material should be preserved in liquid preservative. Kill in a preservative (e.g. spirit, petrol or 'Shell Odourless Carrier' (see Guillarmod 1976)) or put in very hot water for a few minutes, alternatively dry quickly using metal corrugates. See DeWolf 1968: 97; Fosberg & Sachet 1965: 126–127; Jain & Rao 1977: 39–40; Sánchez Mejorada 1986; Smith 1971: 20–23. For the use of microwaves in the preparation of succulents see Fuller & Barbe 1981 and Leuwenberger 1982. (See also below, *Aloe*, *Cactaceae* and *Euphorbia*).

- **Floating or submerged aquatics**. These can be successfully collected by 'floating out' the specimens onto the paper folder, which is itself submerged under the specimen and drawn out of the water at or near the collection site – or by using a wide water-filled receptacle on return to base camp. If the latter, it is important to keep the plants fully hydrated until they can be treated. See Fosberg & Sachet 1965: 107–109; Jain & Rao 1977: 40–41, fig. 14; Lot 1986, Smith 1971: 18–19, fig. 12; Raynal-Roques 1980; Taylor 1977 & 1988; Womersley 1981: 98–99.

- **Cushion or tufted plants**. It is often impossible to press the whole plant. Segments of a suitable size can be pressed, but care must be taken not to fragment them unduly. Whole cushions can be dried in a padded press or box.

- **Large**. If a leaf or stem, subdivide and label sequentially all of the parts. Usually the base, middle and apex is adequate, accompanied by measurement and/or sketches and photographs. For trunks, at least a bark sample should be collected. (See below, *Araceae*, *Musaceae* & *Palmae* and **32. The collection of pteridophytes**). See Jain & Rao 1977: 42; Smith 1971: 19–20.

- **Plants with delicate flowers**. Many short-lived flowers are deliquescent and will disintegrate unless picked early in the day and immediately pressed or placed in spirit. The corollas of several types of flower (e.g. *Iris* and *Hibiscus*) will stick to the specimen folder and easily become damaged. As a general policy detach some flowers, spread the corolla and press in a separate fold of tissue or non-absorbent toilet paper. Do not try to open this folder until the flowers are fully dry. Preserve additional flowers in plastic bottles of liquid preservative (Smith 1971: 23). Fill the bottles completely with liquid; air bubbles can damage delicate material.

- **Complex or obscure structures**. Try to preserve at least one item in liquid preservative.

- **Cauliflorous or ramiflorous plants**. These have the flowers and fruits borne on the trunks or main branches. If possible, detach together with the supporting area of bark. Other plants with flowers or fruits that are difficult to press while attached to the stem – detach and make an explicit note on how they were attached.

- **Plants with bulbs or corms**. Dig up underground parts carefully and remove soil. Small bulbs or corms can be cut in half lengthwise; larger ones should be sliced. These organs will remain alive in the press unless killed (see above, *Succulent or fleshy*).

List of selected families of flowering plants, with short notes on the especial importance of collecting (or recording) particular features.

It is assumed that flowering material is essential and that sterile material by itself is generally worthless. However, a few groups, e.g. *Gramineae – Bambusoideae* (Bamboos) and *Lemnaceae*, are so rarely fertile that their taxonomy has come to be based on sterile material.

Acanthaceae — Fruit, seeds in packet. Dry some flowers separately (cut open).

Agavaceae — (Sánchez Mejorada 1986: 107).

Aloaceae

Aloe — Upper leaf surface with margin (slice off lower surface and scrape away inner tissues). Inflorescence – kill quickly. Record habit, leaf spotted or uniform, inflorescence branching and flower colour.

Araceae — Keep leaf-sheath in one piece if possible and retain ligules at petiole base. Record colour of exudate and if caustic. It is very important to note the presence or absence of **trichosclereids**; these are long white strands which can be observed inside the broken petiole or lamina only while the material is still fresh. (Croat 1985; Nicolson in Fosberg & Sachet 1965 & in Womersley 1981: 132–137).

Araliaceae — Include juvenile foliage.

Aristolochiaceae — Flowers in spirit. Some species are cauliflorous.

Asclepiadaceae — Flowers in spirit. Dig for tuberous roots. If follicles are likely to dehisce during drying tie a thread around them to prevent loss of seeds.

Balanophoraceae — Whole plants in spirit. Sometimes dioecious. Tuber surface important. Parasitic: collect or record host.

Balsaminaceae — Flowers in spirit and some dissected on thin paper or toilet tissue. Ripe seed in packets.

Impatiens — (Grey-Wilson 1980).

Begoniaceae — Male and female flowers essential (may appear on same plant at different times), some in spirit and some dried separately. Collect young ovaries and dehiscing anthers and dehiscing fruit. Collect or note rootstock. Dry quickly.

Bignoniaceae — Cut open and dry some flowers.

Bombacaceae — If spirit not available, take longitudinal sections of flower.

Bromeliaceae — (Aguirre León 1986: 118–119; Smith 1971: 23–24).

Burmanniaceae — Whole plants in spirit, collecting as much of the underground parts as possible.

Burseraceae — May be dioecious.

Cactaceae — Flowers, fruits and whole stems (dwarf species only) of cacti are best placed in spirit if their structure is to be adequately preserved. In the field, treatment of the vegetative parts of cacti for the herbarium depends on whether it is feasible to transport plants back to laboratory conditions in the live state, or whether they must be prepared for drying *in situ*. In the former case they should be wrapped in newspaper and kept dry until the time for preparation arrives. Then, depending on shape, stems will need to be appropriately sectioned, to show shape, number of ribs etc., and most of the fleshy parenchymatous tissue removed with a sharp knife and scapel. If available, a short spell in a microwave oven can facilitate subsequent drying by quickly killing the remaining tissues. The external remains and woody structures can then be dried between cardboard sheets (flimsies and/or drying paper are not necessary), the cardboard sheets are separated by aluminium corrugates and placed in a moderately tightened press in a hot ventilated oven (the temperature should be higher than for conventional use). Once drying is well advanced the press should be tightened with care if the now fragile 'exoskeletons' are not to break.

Preparation for drying in the field is similar, but the application of copious amounts of cooking salt to the cut stem surfaces has been found to greatly speed up the killing and water-extraction process. Salt-coated materials are placed in the press between generous layers of newspaper for 24 hours, after which the brine-soaked paper must be discarded (do not re-use!) before drying between cardboard/aluminium sandwiches is attempted. Presses containing salted cactus materials are best isolated from other material being dried. It should also be mentioned that material dried by the salt method which is destined to be conserved in herbaria in the humid tropics without air-conditioning may be hygroscopic and become damp during storage. The salty material, however, is well protected from insect attack.

Photographs including centimetre scales are an essential part of the preparation of cactus materials for the herbarium and often allow for more rapid identification. See above, *Succulent or fleshy*. (Sánchez Mejorada 1986: 106–107).

Capparaceae — Mature fruiting material. Seeds. Flowers and fruits from same individual plant.

Caprifoliaceae — Ripe fruit.

Caryophyllaceae — Mature fruits and seeds.

Casuarinaceae — Dioecious or monoecious. Ripe fruits.

Commelinaceae — Often flower for a short time during the day; record periodicity if possible. Colour of anthers. Flowers in spirit and/or carefully pressed. Fruits important. Dig up roots.

Compositae — Record size and shape of involucre of capitulum, and preserve some capitula in spirit. Try to collect mature fruiting capitula as well as flowering ones. Press capitula to show upper and lower surface of head. Underground parts.

Coniferae — Although many taxa or organs are amenable to conventional pressing, others tend to fragment on drying. The leaves of *Picea* and *Tsuga* fall off, and the male cones of most species break up, as do the female cones of *Abies*, *Araucaria* and *Cedrus*. Such examples should be immersed in 70% ethyl alcohol as soon after collection as possible, followed immediately by immersion in 50%

aqueous glycerol for four days. They should then be rinsed in water and pressed and dried in the usual way, although separate cones are not pressed. Some untreated fragments should be placed in a packet, and marked as such, in order to preserve the 'bloom' (glaucescence), waxes and other substances soluble in alcohol which would otherwise be lost (for a detailed account see Page 1979).

Convolvulaceae — Split open and press some flowers separately. Ripe fruit truly indehiscent or not? Dig up tuberous roots.

Crassulaceae — See above, *Succulent or fleshy*. (Sánchez Mejorada 1986: 107).

Cruciferae — Mature fruits.

Cucurbitaceae — Usually unisexual, often dioecious; correlated male and female collections are very valuable. The fleshy fruits dry badly, so spirit material plus coloured photographs important. Collect fruit-stalk complete with fruit. Flowers in spirit desirable. Note habit, rootstock, whether wild or cultivated. Collect mature seeds.

Cyperaceae — Ripe fruits (put some in packets). Underground parts. Very young inflorescences of little value.

Dilleniaceae — Ripe fruits.

Dioscoreaceae — Male and female inflorescences, some in spirit. Record position of inflorescence and position of flowers in relation to inflorescence axis. Axillary bulbils. Dig up tubers carefully and describe and photograph. Record direction of twining (clock-wise or anti-clockwise when seen from above). Range of leaves to show size differences and different positions. Fruits and seeds dry.

Dipterocarpaceae — Ripe fruit.

Droseraceae — Several closely allied species may grow together.

Ebenaceae — Dioecious; collect male and female. Ripe fruits with calyx attached.

Elaeocarpaceae — Fruit colour, texture. Aril colour, texture. Seed colour, coat texture.

Ericaceae — Ripe fruits, noting any 'bloom'. Dry quickly to prevent leaves dropping (and/or dip in spirit first).

Eriocaulaceae — Collect inflorescences in different stages. Look carefully for similar species growing together.

Euphorbiaceae — Beware of dangerous sap in some taxa and stinging hairs in *Tragia*. Male and female flowers.

Euphorbia — Herbaceous or woody species. The characters are mostly in the glands, fruits and seeds. Ensure that glands are not obscured by the cyathophylls (leaf-like structures) being folded over them. Flowering material by itself is not much use. Fruiting and seeding material from the same population must be collected (if necessary at different times).

Succulent species – record habit, height; colour slides desirable. Preserve the stem apex in spirit if in flower or fruit. To dry specimens, kill quickly, preferably in petrol or spirit, then place in press. After a few days the limp stems can be sliced longitudinally. *Extreme caution* should be exercised to avoid latex coming in contact with the face, cuts or scratches, or other tender tissues. Wash hands very thoroughly. Petrol is a good solvent.

Gentianaceae — Look out for heterostyly, range of flower size. Collect roots (to determine whether annual, biennial or perennial).

Geraniaceae — Ripe fruit.

Gesneriaceae — Ripe fruit. Some flowers in spirit.

Gnetaceae — Monoecious or dioecious. Male and female flowers and ripe fruit.

Gramineae — The whole plant, including underground parts. Inflorescences should not be too young.
Bamboos — (Koch 1986; McClure in Fosberg & Sachet 1965 & in Womersley 1981: 127–131; Soderstrom & Young 1983).

Hydnoraceae — Spirit.

Hydrocharitaceae — Both sexes needed. Inflorescences in spirit.

Juncaceae — Mature fruit. Underground parts.

Labiatae — Record smell of crushed leaves.
Plectranthus — Leaves will usually fall if plant is not killed before pressing.

Leguminosae — Close-up photographs of flowers and fruits. If collecting for germination place in muslin or paper bags and dry naturally – no forced heat or storage in plastic bags.
Mucuna — beware of irritant hairs.
For ***Mimosoideae*** — collect flowers and fruits.

Lemnaceae — Float onto paper *plus* preserve in spirit (Haynes 1984; Jain & Rao 1977: 40).

Lentibulariaceae
Utricularia — Best placed in liquid preservative. Terrestrial species should be carefully dug up with a knife point; keep the subterranean organs attached if possible, and wash off soil. Diffuse aquatic species can be floated out. Many species are gregarious, so collecting and note-taking must be carried out with great care (Taylor 1977 & 1988).

Loranthaceae — Fruiting specimens alone are useless. Record host plant. Kill then dry quickly to prevent leaf fall.

Marantaceae — as for *Zingiberaceae*.

Melastomataceae — Dry some flowers separately. Petals of many drop early.

Menispermaceae — Dioecious. Fruits important.

Moraceae — Dry fruits quickly. Some inflorescences in spirit.

Musaceae

Musa (banana) — (Fosberg & Sachet 1965: 109–110; Womersley 1976 & 1981: 100–102).

Myristicaceae — Dioecious. Ripe fruits important.

Nymphaeaceae — Look out for variation of flower colour within a population; does flower size depend on habitat? Ripe fruit important.

Ochnaceae — Collect flowers and fruit from the same plant. Some fruit in spirit. Beware of closely allied species growing together.

Orchidaceae — Flowers in spirit very important. Cut longitudinal sections of thick pseudobulbs and tubers of terrestrial orchids; killing may be necessary. Fruit not essential. Colour slides desirable. (Aguirre León 1986: 114–117).

Orobanchaceae — Ripe fruits important. Parasitic: record host if possible.

Palmae — Care should be taken that the hastula (see fig. 48) at the base of many fan-leaved palms (i.e. palmate and costa-palmate palms) is not damaged. If it is necessary to remove part of the lamina, cut it away from one side so as to retain the hastula in one piece. (Balick 1989; Dransfield 1986; Quero 1986; Tomlinson in Fosberg & Sachet 1965 & in Womersley 1981: 119–126).

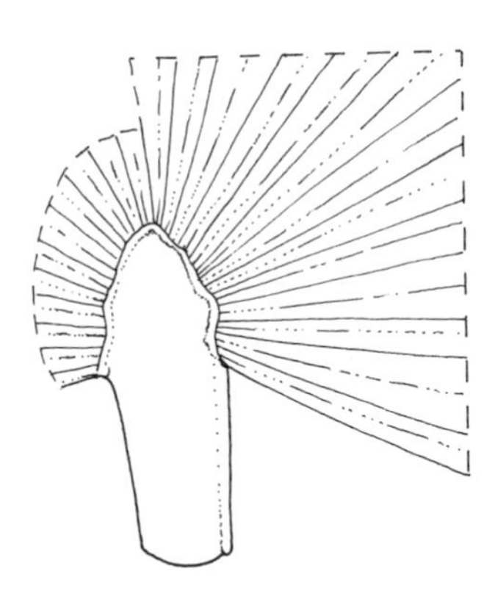

FIG. 48. — Palm leaf trimmed around the hastula.

Rattans – (Dransfield 1979).

Pandanaceae – (Stone 1983 and in Womersley 1981: 94–97).

Passifloraceae – Look for both sexes. Flower and fruit in spirit. Variation in leaf-shape.
Passiflora – (Jørgensen et al. 1984).

Piperaceae – Ripe fruits.

Podostemaceae – On rocks in rivers, flowering as water level falls. Collect whole plants in various stages. It may be necessary to chip away a portion of the rock. Several species may occur together.

Polygonaceae – Fruits essential.

Portulacaceae – Mature fruits and seeds. Some flowers in spirit.

Ranunculaceae – Underground parts. Fruits important.

Rosaceae – Fruits important. Leaves from flowering shoots and sterile shoots.

Rubiaceae – Fruits important. Young and mature stipules. Often heterostylous, sometimes dioecious. Sample so as to indicate branching pattern (see fig. 47 A & C).

Scrophulariaceae – Some flowers in spirit. Record hosts of parasites and hemiparasites. Plants frequently blacken on drying, so record flower colour.

Umbelliferae – Ripe fruits essential. Collect basal leaves, and underground parts if possible.

Urticaceae – Beware of stinging hairs. Ripe fruit. Base and roots. Smaller species are easily overlooked.

Zingiberaceae – Underground parts; inflorescences in spirit. (Burtt & Smith 1976).

- **Bark samples**. Use a chisel and hammer to remove if necessary. Strips up to 10 × 4 cm., or sufficient to show the general appearance may be detached and dried as flat as possible (they tend to curl on drying) to accompany herbarium specimens (see also below, *Wood samples*). Try to cut obliquely across one end to show the number of layers and thickness of each. Remember to record this information as well as colour, colour changes, smell and exudates. Write the collection number directly on the inside of the sample or attach a number tag. (See Matthew 1981: 77).

If samples are required for phytochemical analysis they should be dried away from direct sunlight and heat. Pieces of bark 1.5 cm.[3] with some wood if possible, may be needed for anatomical study; they should be preserved in 70% alcohol (see **31. Collecting materials for ancillary disciplines** and ter Welle in Campbell & Hammond 1989: 467).

– **Wood samples**. A voucher specimen with foliage and flowers or fruit should always be collected to accompany a wood sample; it is important that sucker or coppice shoots are not used for this purpose. Avoid taking wood samples from diseased plants with active fungal growth and from buttresses and branches, where the wood structure will be atypical. The sample should include adult wood.

Samples of the trunk may be obtained from felled trees (trees should not be felled especially for this purpose) or by cutting a piece from a standing tree. This is a skilled technique, best carried out as follows (see fig. 49):

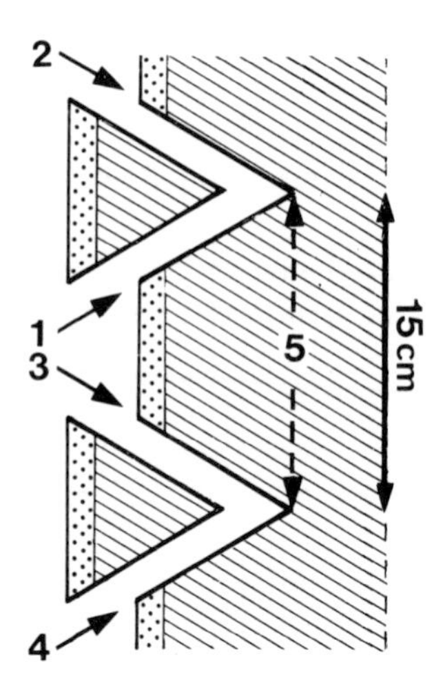

FIG. 49 — Sequence of cuts, 1–5, for removal of a wood sample.

– remove a wedge-shaped segment, making the lower (upwardly angled) cut first and the upper (downwardly angled) cut second.

– at a position ± 15–30 cm. below this wedge, remove a second wedge. This time start at the upper cut and make the lower cut second.

– starting at the upper niche, left by the removal of the first wedge, cut downwards to the lower niche. Remove the sample by freeing either side.

An entire section of the main stem may be taken from small trees or lianes up to 15 cm. in diameter, this should be c. 10 cm. long. If the diameter is greater than 6 cm. the sample can be split lengthwise. If possible, samples from lianes should be taken from both near the ground and the canopy.

The samples should be trimmed as necessary to the *maximum* recommended size of c. 30 × 10 × 7 cm.; however, 12 × 8 × 4 cm., with the thickness including the bark, is preferred by some anatomists. If the bark is notably thick and succulent it may be removed to facilitate drying or occasionally it may loosen or separate spontaneously. Detached bark should be secured to the wood sample with string. The samples should be clearly numbered and left to dry in a shady place, although sunlight and heat from a stove will not cause damage as long as the samples are not dried too quickly. Dampness, however, must be avoided. If the bark is to be used for chemical analysis, heat or direct sunlight should be avoided (see **31. Collecting materials for ancillary disciplines**). Certain families (some Leguminosae, Loganiaceae, Menispermaceae, Nyctaginaceae, Solanaceae and Urticaceae) have a special type of phloem (included or interxylary phloem) which can be destroyed by drying. If required for anatomical study small samples (c. 1.5 cm.^{3}) can be preserved in 70% alcohol. See Stern in Fosberg & Sachet 1965: 127–129; ter Welle in Campbell & Hammond 1989: 467–468; Womersley 1976 and 1981: 121–122.

Collecting tree-canopy material and epiphytes

Many epiphytes as well as parts from trees and lianes fall from the canopy during heavy rain, strong winds, or the passage of arboreal animals. Examine the ground below for fallen epiphytes, twigs, branches, flowers and fruits. If any organs are gathered from the ground try to find out where they came from and note on the field data, 'picked up from the ground', or 'fallen flowers, fruits etc.'. Newly fallen (or felled) trees are not uncommon, especially along new roads. Host and epiphyte data are all too rare. Try to get the host tree identified; an experienced local botanist may be able to name good, sterile material.

Since a large proportion of vascular plants will be beyond the reach of a long-handled pruner or pole, the following means of collecting them can be considered:

- ***Tree-climbers***
  With the services of skilled tree-climbers much can be obtained. Only they can decide how best to reach the material.

- ***Tree-climbing techniques***
  There are various techniques for reaching tree canopies. All require perseverance and skill to master and are best explained by and practised with skilled practitioners. The following references are useful:

  Hyland (1972)
  Kuhlmann (1947)
  Mitchell (1982)

Wendt (1986)
Womersley (1981)

- ***Other techniques***
  Shooting a weight with a light line over a branch and then hauling up a thicker line can be a useful means of pulling down twigs and small branches. It is also the first stage of the preparations for the 'free-climbing' technique.

Numbering the Collections

Each collection must have a number. The collection can be made up from parts of one large plant or smaller individuals from one population, collected in the same locality and at the same time by one collector (or group of collectors).

It is not advisable to use the same number more than once. If collecting from a marked individual tree, where flowers and fruits are being sought over a period of months, separate numbers should be used for each collection, cross-referenced to each other.

Each number should be a unique number from the collector's single running-number sequence, which should start at 1 and continue ± unbroken. Numbering within each year (e.g. *F. Bloggs* 1/87, 2/87 then 1/88, 2/88 etc.) or for each locality or country (e.g. *F. Bloggs* K1, K2, say for Kenya) may be of use to the individual collector, but can easily be misinterpreted by a botanist who may wish to cite the collections. Complicated numbers or codes with reference to day, month and even locality are occasionally used, but these should be avoided, especially because they cannot usually be accommodated in computer databases.

In pencil, record the collecting number on a tag attached to each or most of the specimens; it is also helpful to write it on each flimsy (folder). For plants (or organs) that have been subdivided, tags are essential for clearly indicating which part goes with which, and in what sequence.

If after naming, a single numbered collection, considered in the field to be one taxon, is ultimately found to be a mixture of two or more taxa, these should be designated by suffixes A, B, etc. to the number. Where a small plant was inadvertently collected in addition to the main specimen (i.e. the data only applies to the main specimen), add an A only to the additional collection.

RECORDING DATA

Equipment required for recording data

- *Collecting books.* As sturdy as possible and preferably bound, as this provides a firm writing surface. Pocket-sized is most convenient. Many herbaria design and produce notebooks to their own specification.

– *Diary or notebooks.* For recording itinerary, topography, people met, travel details, field sketchs etc. Such books can be invaluable in subsequent dealings with the collections. If collecting books are lost, which sometimes happens, at least some data can be salvaged.

– *Pencils and pens.* HB–2B are the best; brightly coloured ones are less likely to be lost. Propelling pencils are convenient as they will not need sharpening, but take spare leads. Ball-point pens may also be useful but do not use pens with ink that is not water-resistant.

– *Tape recorder* (for personal dictation). Not essential, but can save time in the field. It must be reliable and supplied with an adequate number of batteries.

– *Portable (lap-top or notebook) computer.* Small light-weight models with solar batteries make the field use of computers feasible. However, check local

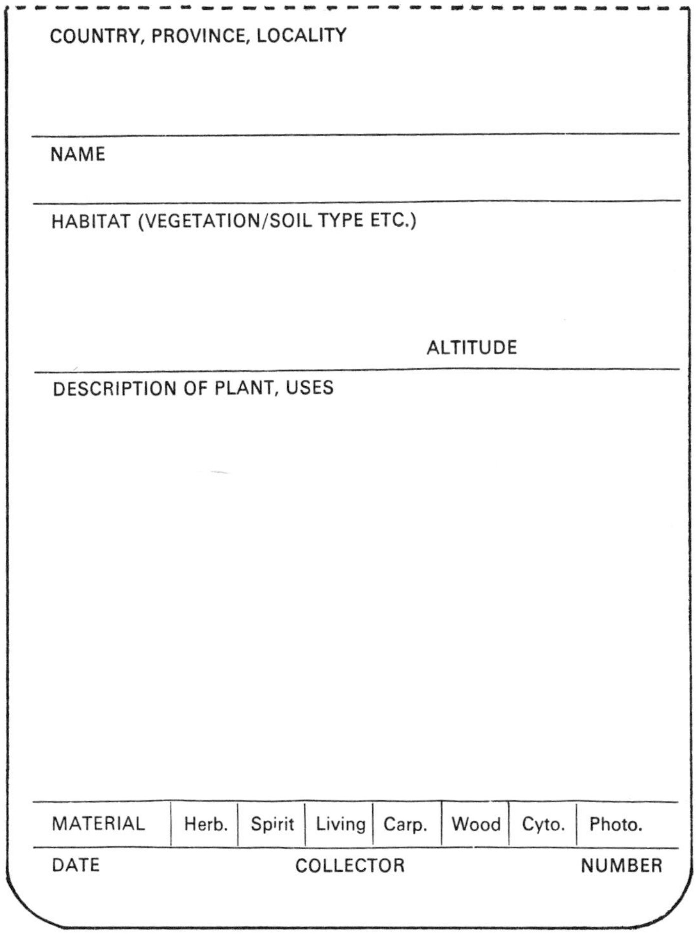

COUNTRY, PROVINCE, LOCALITY

NAME

HABITAT (VEGETATION/SOIL TYPE ETC.)

ALTITUDE

DESCRIPTION OF PLANT, USES

MATERIAL	Herb.	Spirit	Living	Carp.	Wood	Cyto.	Photo.

DATE COLLECTOR NUMBER

FIG. 50. — Page from printed collector's notebook.

customs regulations before taking one into another country. Some written back-up of field data is still recommended.

- *Photographic equipment.* See **38. Photography and field-work**.

- *Altimeter.* Remember to reset.

- *Clinometer*, or '*Haga Altimeter*' or '*Blume-Leiss Altimeter*' for measuring the height of trees.

- *Tape measure* and length of rope for measuring the diameter of tree trunks.

- *Hand-lens.*

Record field data completely and methodically, preferably *at the time of collection.* Experienced collectors usually record all relevant data in ± the same sequence and position in their notebooks, in neat handwriting, usually in pencil. The production of herbarium labels (and photocopies of the notebooks for safe-keeping) is greatly simplified if this is done. Specially produced notebooks with printed headings for the basic data are very useful (fig. 50).

The final production of data labels is dealt with in **6. Label design and production** (see fig. 51 and 52 and below, *Selection of duplicate sets*).

The minimum data to record for any collection are:

- Locality, including the country.
- Habitat, altitude.
- Description of plant(s).
- Collector's name and number and date of collection.

Anything less is bad practice. Also put in a field identification, even if it is only to family or group, e.g. Fern.

Full data

Locality

- This should be given as precisely as possible, stating the longitude and latitude or other grid reference. A printed sketch map marking the locality can eventually be put with the specimen (see fig. 52).

Habitat and ecology

- Vegetation type, associations with other species (list adjacent species and/or collection numbers), life-forms. (See **40. Ecology and the Herbarium**).

- Relative abundance: how many, how common? Growing gregariously or singly.
- Variaton of any feature(s) of the population.

Habit

- Tree (including overall shape (see Whitmore 1972)), shrub, climber, epiphyte (and type of epiphytic habit), herb etc.

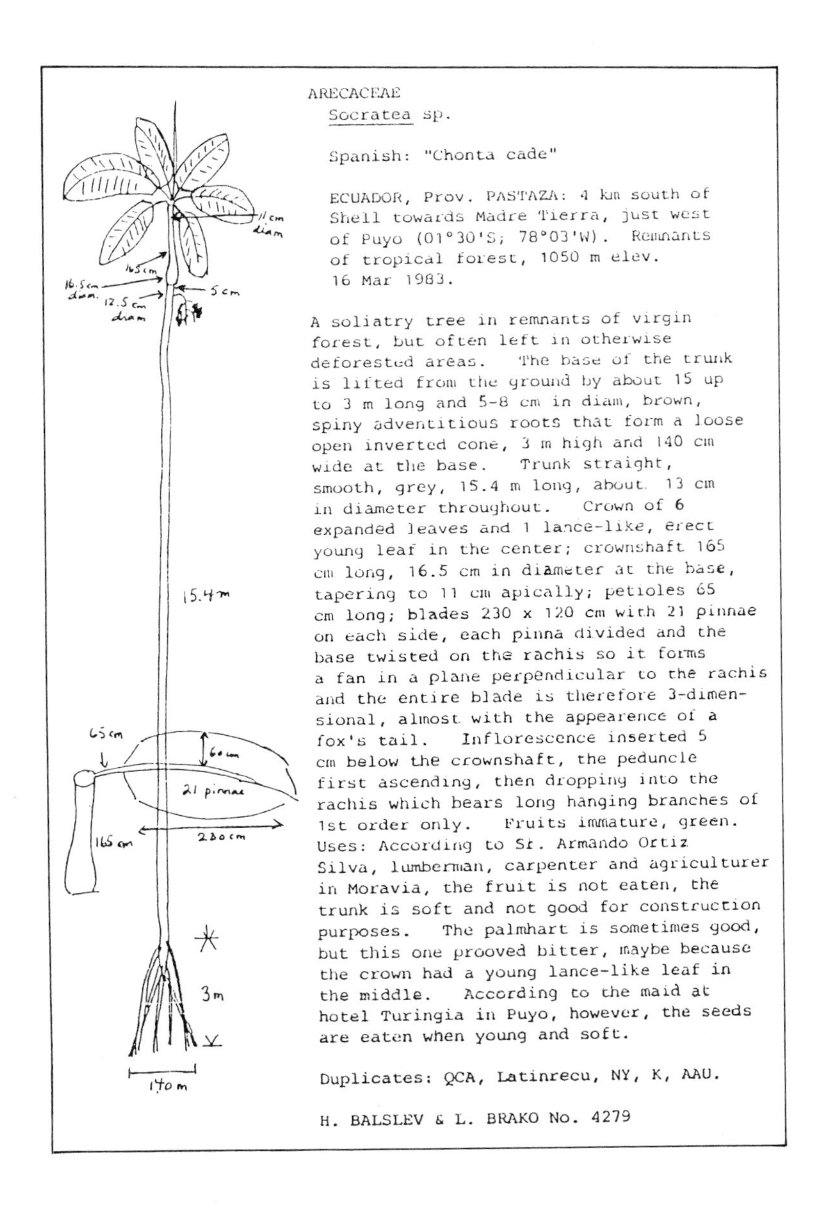

ARECACEAE

Socratea sp.

Spanish: "Chonta cade"

ECUADOR, Prov. PASTAZA: 4 km south of Shell towards Madre Tierra, just west of Puyo (01°30'S; 78°03'W). Remnants of tropical forest, 1050 m elev. 16 Mar 1983.

A soliatry tree in remnants of virgin forest, but often left in otherwise deforested areas. The base of the trunk is lifted from the ground by about 15 up to 3 m long and 5-8 cm in diam, brown, spiny adventitious roots that form a loose open inverted cone, 3 m high and 140 cm wide at the base. Trunk straight, smooth, grey, 15.4 m long, about 13 cm in diameter throughout. Crown of 6 expanded leaves and 1 lance-like, erect young leaf in the center; crownshaft 165 cm long, 16.5 cm in diameter at the base, tapering to 11 cm apically; petioles 65 cm long; blades 230 x 120 cm with 21 pinnae on each side, each pinna divided and the base twisted on the rachis so it forms a fan in a plane perpendicular to the rachis and the entire blade is therefore 3-dimensional, almost with the appearence of a fox's tail. Inflorescence inserted 5 cm below the crownshaft, the peduncle first ascending, then dropping into the rachis which bears long hanging branches of 1st order only. Fruits immature, green. Uses: According to Sr. Armando Ortiz Silva, lumberman, carpenter and agriculturer in Moravia, the fruit is not eaten, the trunk is soft and not good for construction purposes. The palmhart is sometimes good, but this one prooved bitter, maybe because the crown had a young lance-like leaf in the middle. According to the maid at hotel Turingia in Puyo, however, the seeds are eaten when young and soft.

Duplicates: QCA, Latinrecu, NY, K, AAU.

H. BALSLEV & L. BRAKO No. 4279

FIG. 51 — Example of full data.

Underground parts

- Tap root, fibrous roots, tubers on roots, extent of roots.
- Scent of cut parts.
- Rhizome – depth in soil, length, spacing of shoots.
- Bulb, corm or tuber – size and shape.

FLORA OF BRAZIL

Territory of Roraima

No. M. 184 Rubiaceae

SEMA Ecological Reserve, Ilha de Maracá, Roraima.

Low forest with abundant palms close to a seasonal lake. 3°23'N 61°27'W.

Slender single-stem shrub, 3 m. Stem green, becoming reddish towards the top. Pedicels and peduncle orange. Corollas orange with deep purple tips. Young fruits green.

Date. 6 5 1987

Coll.: William Milliken

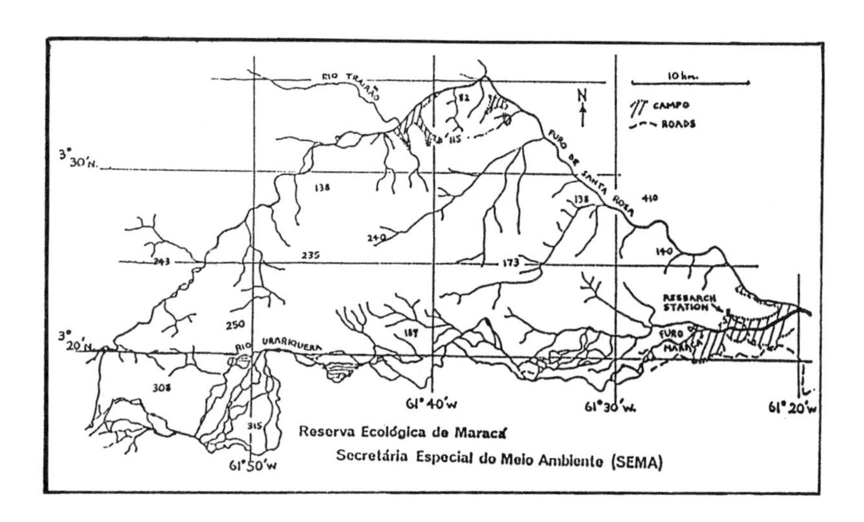

FIG. 52 — Example of good data with printed sketch map.

Stems and trunks

- Size: total height, girth at breast height or diameter at breast height (G.B.H./D.B.H.) (estimated or measured).
- Height of trunk or stem before branches; whether buttressed or not.
- Bark colour, texture, thickness, lenticel colour.
- Wood hardness, colour, grain type.
- Diameter at various heights (if variable).
- Cut trunk ('slash'), sap or latex including colour, smell, consistency and other properties.
- Shape in cross-section (circular, fluted etc.).
- Internode lengths.
- Thorns, spines, especially if on trunk.

Leaves

- Deciduous or evergreen.
- Texture, colour(s), smell, glossiness, glaucescence.
- Exudate or glands.
- Orientation in relation to petiole or stem etc., e.g. 'pendulous', 'horizontal'.
- Large and/or compound (collect sequentially and label each part).
- Outline (if large or complex), sketch.
- Note if heterophyllous (juvenile, shade or submerged aquatic leaves).

Inflorescence

- Exudate or glands.
- Cauliflorous, ramiflorous, any other data on position or form that may be lost in prepared specimen.
- Colour of axis.

Flowers

- Note if heterostylous or if plant monoecious or dioecious.
- Scent.
- Corolla colour, texture.
- Calyx colour, texture.
- Exudate or glands.
- Behaviour (e.g. open early and closed by 12 noon).
- Pollinators.

Fruit and Seeds

- Smell.
- Colour, texture.
- Size, shape.
- Seed-coat colour, texture.
- Aril colour, texture.
- Dispersal (animals, wind or water).

Local name

- Record name phonetically if the correct spelling is uncertain; try to check reliability of name.

Uses

- Record any uses; try to get confirmation (see **39. Economic botany and the herbarium**).

PROCESSING SPECIMENS

Equipment needed for processing and packing specimens

- *Scissors & secateurs* for trimming specimens.

- *Double flimsy*, ('specimen folder') a folded sheet of thin strong paper within which each plant (or portion of a plant) is retained throughout the drying process. The flimsy should be the same size as the required herbarium sheet. (see also **5. Materials**).

- *Drying paper*, special drying paper is excellent but can be very expensive.

- *Absorbent cloth*, zig-zag folds of inexpensive cloth such as sacking or scrim are preferred to drying paper by some collectors using the Polish press method (see below, *Polish presses, or Madalski's method of drying plants*). It is more convenient to have pads made but the advantages in using cloth are minimal. Also folds of muslin for sticky plants (see below, p. 218) and for spirit (see **31. Collecting materials for ancillary disciplines**, *Collecting in liquid preservatives*).

- *Newspaper*, this can generally be substituted for both flimsies and drying paper (see also **5. Materials**).

- *Press frames*, these should be slightly larger than the flimsies and drying paper and no bigger than the mounting paper. A pair is needed for each press. Press frames must be latticed; they may be made of wood, metal, bamboo or rattan, strong and as light-weight as possible. It is possible to purchase ready-made presses (both metal and wood) but they are cheaper to make. Ensure that all rivets, screws etc. are rustproof. See fig. 57.

Polish presses (see p. 222, *Polish presses, or Madalski's method of drying plants*) are made of welded steel mesh; a 12 mm^2 mesh is ideal. If this is cut from a roll the edges should be filed smooth and bound with a tough protective tape for ease of handling.

- *Press straps*, strength and ease of applying and maintaining pressure are the requirements. These can be improvised from many materials. If there is a choice, use cloth webbing straps with spiked buckles. Any plastic (including nylon) can melt.

- *Cord*, preferably terylene, for securing Polish presses (see fig. 58). Each press will require about c. 2 m. length.

- *Corrugated sheets.* (corrugates), preferably made of aluminium. Cardboard is less effective; the corrugations should run widthwise. These speed up the drying of specimens by ventilating the press and conducting heat. Not easy to improvise although lattices or sewn 'blinds' of bamboo or wooden slats are useable.

- *Cardboard supports*, slightly larger than the specimens – used to support specimens collected by the alcohol method or for the final bundles.

- *Paper slit-slips*, strips of any fairly stiff paper (bond or cartridge) about 8 × 2 cm. with a central slit along the length to within 1 cm. of each end. Used to secure bent stems when arranging the specimen (see fig. 56B).

- *Toilet tissue*, non absorbent or thin paper for protecting delicate flowers.

- *Paper or mesh bags*, for carpological material.

- *Plastic bottles*, for collecting separate organs and portions into liquid preservatives.

- *Polythene bags*, big enough to contain whole bundles of dry or spirit specimens. Heavy gauge (500+) polythene bags or tubing are invaluable for wrapping bundles of specimens prior to transport. (see below, *Wet (spirit) collecting*).

- *Adhesive tape* (plastic, linen or paper) and *string*. Not 'Sellotape' or similar (too impermanent and eventually messy). For sealing specimen bags and as specimen number labels. Ideally the gum should not be alcohol-soluble. Back up by also securing with string.

- *Field stoves* and *fuel* for heat-drying, preferably of the type(s) familiar to the collector. Usually the heat source is an oil, kerosene or 'Gaz' lamp (see Plate

1, D–F), but alternatives can be improvised with a trench and glowing embers (see fig. 53). Some means of securely supporting or suspending the press vertically over the heat sources is needed, and if possible shuttering should be placed around the heat source to create an up-draft of warm air. Purpose-made wooden or aluminium frames fitting together to form an open-topped box will fulfil both purposes. Shuttering can be improvised with aluminium corrugates, wire mesh or net or fireproof fibreglass curtains. See Botha & Coetzee 1976; Hallé 1961; MacDaniels 1930 & Schnell 1960: 26–27; Womersley 1981: 33–35.

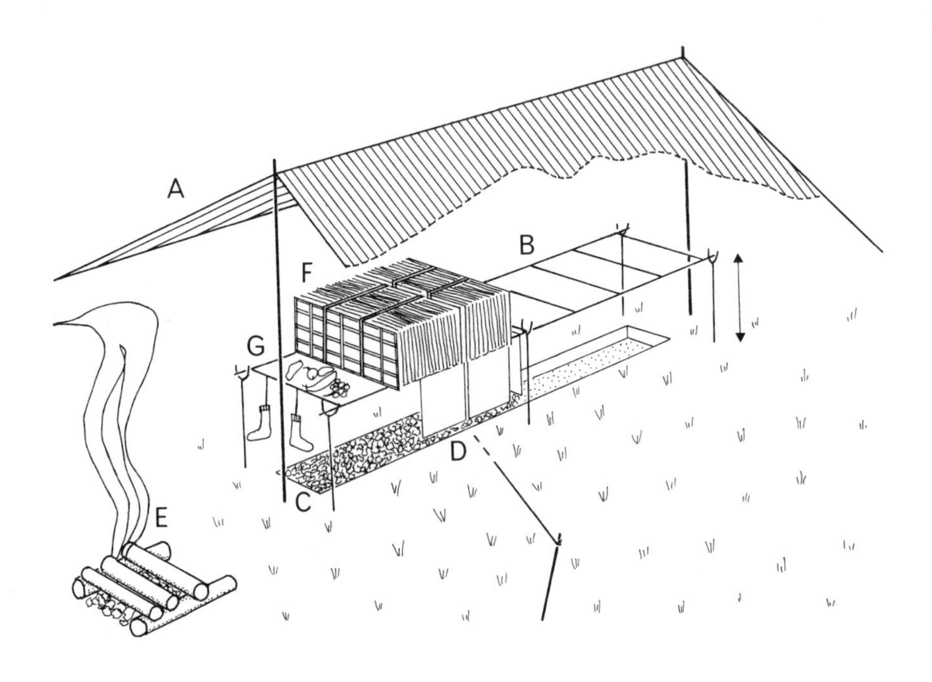

FIG. 53 — Trench and glowing-ember method of drying material. **A** rainproof canopy; **B** supporting frame (if possible use metal cross-bars); **C** trench filled with embers as far along as necessary; **D** skirt of metal corrugates to conserve heat (note gap at bottom); **E** wood or charcoal fire to supply embers; **F** full presses (stacked on sides); **G** large fruits etc. drying on metal corrugate. (N.B. should be checked at least once every 2 hrs; at night no additional embers should be added at least 1 hr before retiring).

- *Wax crayon* and/or *Indelible ink felt-tip pen*, for numbering wood samples. Also for addressing parcels.

- *Strong cardboard boxes* for packing bundles ready for dispatch.

- *Insect-repellant crystals* (naphthalene or paradichlorobenzene). Sprinkle inside polythene bags of dry specimens when packing for transit.

- *Silica gel granules*, sprinkle inside polythene bags of dry specimens.

PLATE 1 — Field work in Brazil: **A** collecting a large fruited specimen; **B** using a field portfolio; **C** interior of tent with bundles of dry specimens packed in polythene bags; **D–F** using a drying stove. Photograph by Andrew McRobb; Kew staff members: **A** Raymond Harley; **B** & **C** Brian Stannard; **D** Eimear Nic Lughadha.

— Tightly sealed bulk containers of alcohol, etc.

— Liquid preservatives (see **31. Collecting material for ancillary disciplines**).

It is important to estimate how much drying equipment should be taken into the field. The number of specimens obtainable may vary depending on the objectives of the trip (general or specialist), the number of duplicates required, the floristic richness of the area and the season. It is important to consider how many people will be collecting, the type of terrain and climate, the proposed method of drying and whether there will be a base camp. Advice should be sought from somebody with experience in the area concerned.

Arranging and preparing the specimens

The specimens should be placed in either flimsies or newspaper. The arrangement is best done when the plants are not too fresh; many collectors prefer to make a quickly arranged pressing overnight and then to carry out fine adjustments the following morning when first changing the drying papers.

— fold and/or trim leaves to fit the sheets as necessary, see fig. 54, A & B.

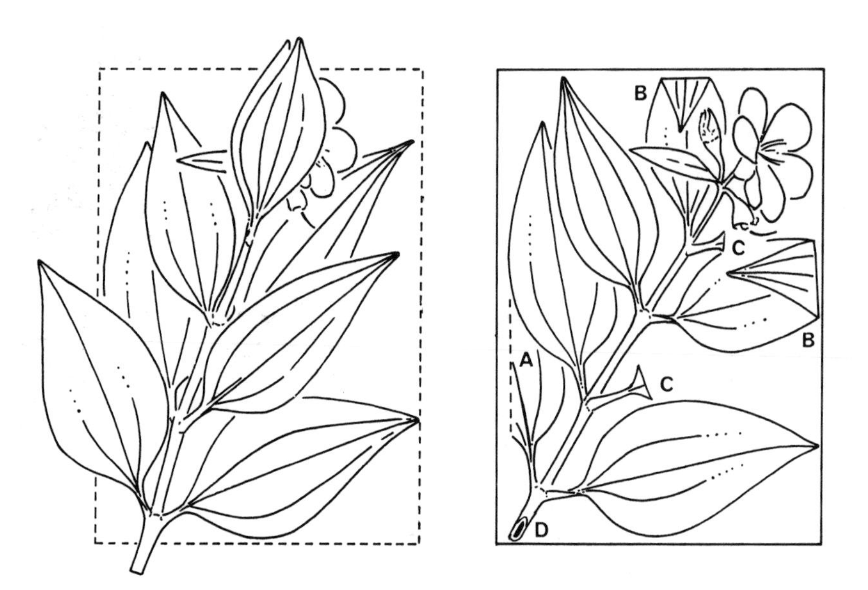

FIG. 54 — Specimen trimmed to fit sheet. **A** trimmed leaf; **B** folded leaves; **C** leaf blade removed but petiole retained; **D** stem trimmed oliquely.

- if leaves are crowded some can be removed but at least the petioles should be left attached to show the position of the leaves (see fig. 54C).

- cut stems and branchlets obliquely to show the internal structure, i.e. whether the stem is hollow or contains pith (fig. 54D).

- large leaves can either be trimmed on one side of the midrib and folded (see fig. 55A) or cut into sections (see fig. 55B).

If the specimens have **thick** or **lumpy** parts, folds of drying paper or newspaper may be added for support or as padding to the more delicate structures to help to distribute the pressure. If this is not done, delicate leaves and petals may receive

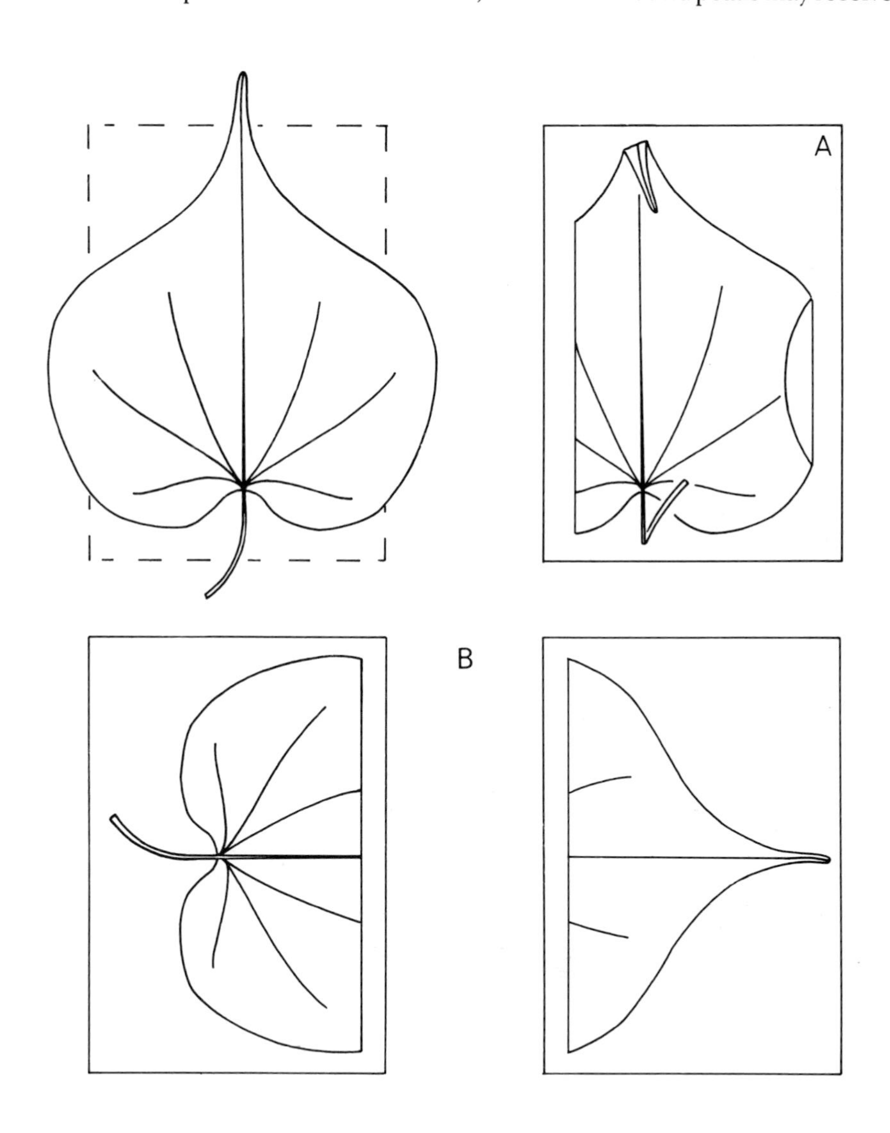

FIG. 55 — Dealing with a large leaf. **A** trim and fold; **B** cut in two sections.

no pressure and therefore dry wrinkled. Thick stems and roots can be cut lengthwise, and projecting branches trimmed off. Drying **leathery (coriaceous)** leaves can take a long time; if they overlap, sandwich a strip of drying paper between them and change it with the drying papers. They can appear dry when they are not – bending them with ease without breaking gives a good indication of inadequate dryness.

If the specimens are **sticky** with mucilage, resin or gums (e.g. *Hibiscus*, *Dodonaea*) they should *not* be placed in newspaper. Either a flimsy or a fold of muslin cloth can be used; when fully dry it can be eased off the material (see Jain & Rao 1977: 41).

The material should be **arranged** so that the maximum information is displayed. To a limited extent poor pressing can be compensated for by careful preparation before mounting (see **9. Mounting herbarium specimens**: 67–73), but such specimens can never equal well and intelligently pressed material. In a well pressed specimen, it will be obvious which side should be mounted uppermost, except in plants such as grasses. The following points should be borne in mind:

- The leaves should be spread so as to avoid as much overlap as possible, at least one should be turned to expose the undersurface, ideally both an old one and a young one. Both aspects of the flowers should be displayed (see fig. 56A).

- Stems should be bent to fit the sheets. If they tend to spring apart they may be restrained in paper slit-slips while in the press (see fig. 56B).

- Any extra flowers collected can be spread out and dried in folds of toilet tissue; if tubular, the corolla should be cut lengthwise and opened out.

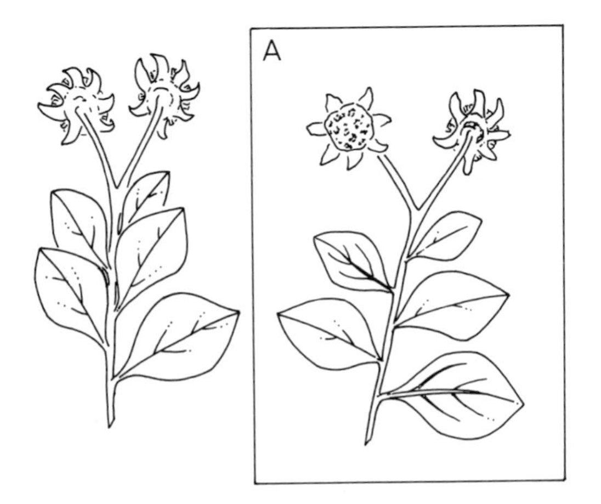

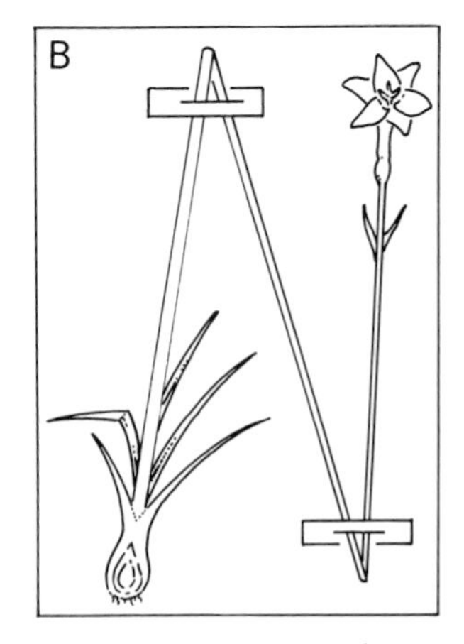

FIG. 56 — **A** show both aspects of leaf and flower; **B** bent stem held in paper slit-slips.

– If there are extra fruits some should be cut longitudinally and some transversely; if large they may be sliced and the individual sections spread out and dried.

Preserving the collections

The specimens will ultimately be dried and pressed flat, except for supplementary carpological or spirit material. There are two basic methods: the specimens may be either directly pressed and dried in the field or collected by the alcohol (or Schweinfurth) method and then pressed and dried at a later date. There are advantages and disadvantages to both methods; often the two can be combined to advantage:

DRYING DIRECTLY IN THE FIELD	WET (ALCOHOL) COLLECTING 'SCHWEINFURTH METHOD'
In humid conditions mould and decay may occur before the plant is dry.	Method especially advantageous in humid conditions as mould will not occur and decay will not start at least for several months.
Time consuming in field, few specimens can be collected.	Saves time in field, more specimens can be collected and then dried at a later date.
A large amount of equipment is needed, presses and drying papers as well as flimsies or newspapers.	Less equipment needed: alcohol, flimsies, or newspapers, cardboard supports and strong polythene bags.
The specimens produced are usually of a better quality; they tend to keep their colours and are not brittle.	The specimens produced often dry blackish or brown and are often brittle. Occasionally flower colour can be quite altered.
Alcohol-soluble substances are not dissolved (e.g. products of glands, waxy cuticles or chemicals, especially flavonoids).	Alcohol-soluble substances can be lost (see Cooper-Driver & Balick 1979 & Coradin & Ginnasi 1980). For this reason it is important to record the use of a solvent on the data label.
It may be necessary to kill certain taxa (e.g. *Loranthaceae* or succulents) before they are put into the press, in order to prevent leaf fall or continued growth.	All plants are killed by the alcohol, and further measures are not necessary.

DRYING DIRECTLY IN THE FIELD	WET (ALCOHOL) COLLECTING 'SCHWEINFURTH METHOD'
Reference to previously collected material is possible in the field.	Reference to previously collected material is impossible in the field (until specimens are dried).
Duplicates are soon available, e.g. for the host country.	Removal of duplicates must be postponed until collection dried.
Once dried, the material must be kept dry in the field.	No such provision necessary.
Material must be securely packed for final transit and kept dry, but there will be no problems with smell.	Packing need not be as strong or waterproof, but the bundles may smell. Ethyl alcohol over 70% concentration is flammable and may be prohibited as air freight.

a. Drying specimens in the field

Place the prepared specimen on an open folder ('flimsy' or newspaper), arranging it as carefully as possible (see p. 216, *Arranging and preparing specimens*). The collection number should be written on a number tag attached to the specimen; it is also helpful to write it on the folder. It is especially important to add numbers to sequentially collected parts of a large plant, i.e. palm or tree-fern fronds. The specimen should stay in its folder until the ultimate destination is reached.

There are two main ways of drying specimens in the field: the conventional method or the Polish press method; often the two can be combined to advantage.

Conventional method

Sandwich the specimen in its folder between sufficient pieces of drying paper (and any padding necessary), then a corrugate (if available) and so on. If few or no corrugates are used, keep the filled press size small, it will dry quicker. With many corrugates the press size can be extended to the limits of the straps. The straps should be fastened as tightly as possible; tighten the first strap moderately firmly, fully tighten the second strap, then return to the first strap and fully tighten. If the straps are fastened in opposite directions the pressure will be more even (see fig. 57 & front cover).

Note that bulky specimens, thick stems etc. can be placed directly against a corrugate to speed up drying.

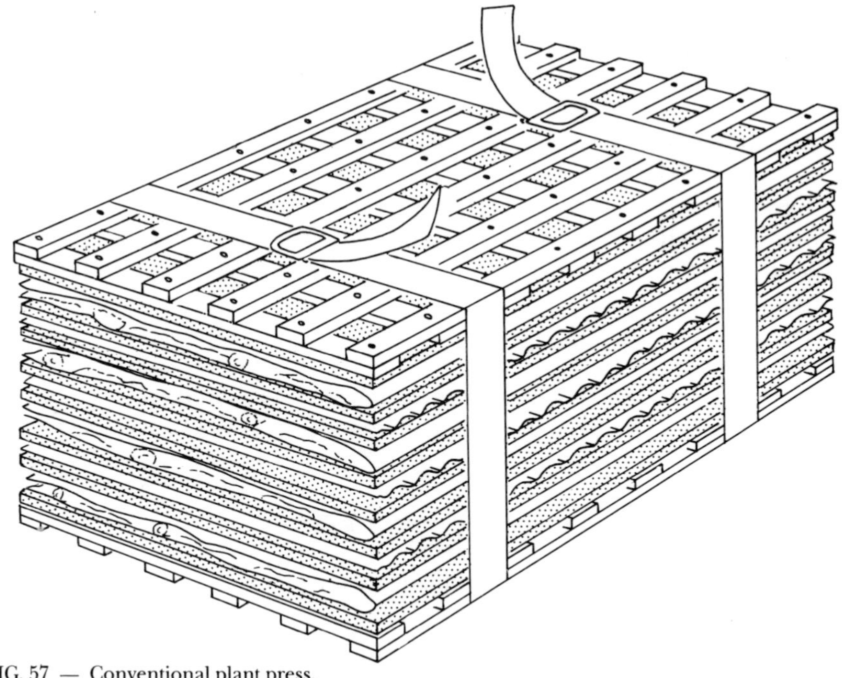

FIG. 57 — Conventional plant press.

Expose the presses to a *gentle* (35–45°C) heat source for as long as is necessary to dry them properly. This does not mean so dry as to be too brittle. Temperature alone is less important than a steady flow of warm air around and (via the corrugates) through the presses. Also excessive initial heat can cause specimens to become 'stewed' (recognised by discoloration and 'cooked cabbage' smell) especially if the drying papers are not frequently changed or the plant is succulent. Some authors warn against rapid drying, but generally it is considered desirable (see Allard 1951 & Camp 1946). The presses (and any loose material) must be well supported so that they cannot fall and start a fire. Make sure that the straps do not hang down and catch fire or the presses char. (See Schnell 1961: 21–32).

Each press should be examined regularly for tightness and turned so that the heating is even. If allowed to become loose, undue distortion or shrivelling will result and items can be lost (add a second folder the opposite way round if necessary). Examine at least twice a day and change the drying papers as necessary. Take advantage of the first change of drying paper to rearrange specimens to best effect, especially flowers. Dry specimens should be removed as soon as possible and the drying papers made ready for re-use. Allow 18 hours to 4 days (or more) for complete drying. The specimens will be rigid when dry, unless very delicate. Not all the material of one number may dry at the same rate so the specimens will need carefully reassociating.

Every portion must bear a collection number. Bulky items for drying can be securely tied on top of the press or between the presses.

Keep the prepared material dry. If the moisture content in the air is high, the material should be placed (after cooling) in polythene, which is then lightly sealed. If available, add silica gel granules. See Plate 1 C.

Alternatives to using artificial heat sources (field stoves or drying cabinets) are:

Sun-drying
This works well if conditions allow. A good rule is to keep the press size small and allow free air circulation. Corrugates are very useful; make several changes of drying paper.

Forced-air drying
Utilizing the roof-rack of a vehicle in motion works well in drier places, but *tie well.* Corrugates are advisable; change drying papers as necessary.

Polish presses or Madalski's method of drying plants

This method relies on keeping the filled presses slim so that all the moisture is lost directly from the surface rather than being removed by the paper changes. It is particularly useful in dry sunny areas, e.g. Mediterranean climates, where it saves the labour of paper changing, and far fewer sheets of drying paper need be

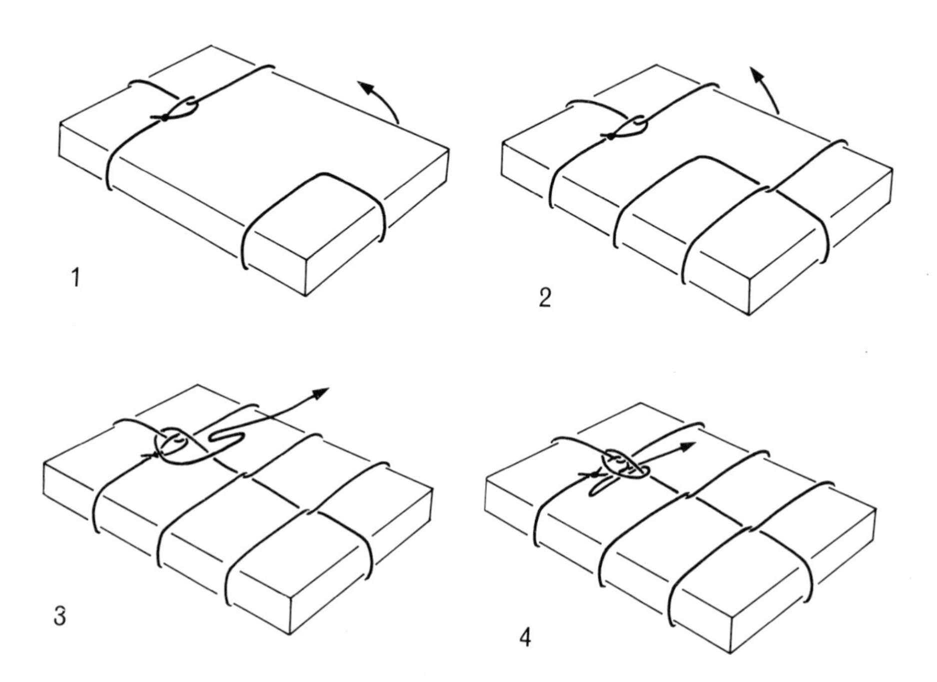

FIG. 58 — The method of tying Polish presses.

carried into the field. It is ideal for plants such as grasses and Ericas that need little arrangement, but is not recommended for complex or succulent subjects.

The prepared specimens are placed in an open folder in the same way as in the conventional method. However, more care is needed in their arrangement as they will be dry before re-inspection. Place in a Polish press (see above, *Equipment needed for processing and packing specimens*) up to 10 folders (fewer if at all succulent) separated from each other and from the press frames by only single sheets of drying paper, absorbent pad or zig-zag folds of absorbent cloth. If the press frames are at all curved, position them with the concave faces outwards, then tie up tightly parcel-fashion with the cord to give even pressure (see fig. 58). No further tightening is required during drying. The presses can be placed in direct sunlight, over a heat source or on the roof-rack of a moving vehicle. It is claimed that in Mediterranean sun the press can get too hot to handle but the contents dry in 3–9 hours. Because of the quicker evaporation, heat is less likely to darken the specimens. (Madalski 1958; Skvortsov 1977, figs. 8 & 9).

b. The alcohol or 'Schweinfurth' method

Specimens (with collection numbers) are placed between folders of newspaper (or in flimsies) only, and when a convenient-sized bundle is reached the whole is compressed and tied, any projecting parts being shortened or suitably padded to prevent puncturing of the bag. Bulky or spiny material is best placed near the centre of the bundle. The bundle is then placed in a heavy gauge (500+) polythene tube with one end folded over three or four times and sealed to make it liquid-proof or into a long bag; 40 cm. wide by c. 140 cm. long. About 0.5 litre (per 12–15 cm. bundle) of 60–80% industrial (or other) alcohol is sprinkled inside. However, a higher concentration (95%) can be used to *kill* resistant tissue (e.g. some Araceae). The open end is then folded and also sealed; several pieces of non-alcohol-soluble tape should preferably be used. The alcohol fills the bag as vapour, and preserves the contents for weeks, *if unpunctured* even up to 6 months. It is important to check the bundles every 4 weeks and, if they show signs of drying, add a cup full of alcohol solution. As many of these bundles as possible should be packed into a suitable strong hessian or woven plastic sack for extra protection. The whole is very robust and can be despatched in this form. See Womersley 1981: 37–39, fig. 13.

Preserving material in spirit

Soft fruits and flowers (or other organs) with a complex structure are best preserved in spirit. If the specimens are to be directly dried, place any items needed in spirit straight away; if the alcohol method is used, all but the most delicate or complex flowers can be transferred to spirit at a later stage.

The method is given in **31. Collecting materials for ancillary disciplines**, *Collecting in liquid preservatives*.

Preserving carpological material

Most fruits can be placed in a paper or mesh bag and air dried. If there is danger of mould developing, treat with alcohol or other fungicide. If possible cut at least one fruit longitudinally and one transversely.

SELECTION OF DUPLICATE SETS

Ideally all the collected material should be returned to the home institution for checking before duplicates are removed. Sometimes, however, one or more complete sets may have to be left in a host country's institution before departure. Such provisions are both usual and courteous, and any requirements should be complied with.

Try to ensure that each collection is divided up into good sets of duplicates; fertile portions should be equally apportioned. (see **8. Processing unmounted materials**). Whenever possible, the collector should personally separate out the duplicate sets. Always ensure that if very few good sets are possible, one at least (the top set) should contain all variations, organs and stages available under that number.

DISPATCH

The specimens should be packed and documented for dispatch to the home institution in a manner similar to that described under **7. Centralized accessioning, recording and dispatch procedures**, *Packing and associated paperwork.*

Once they have arrived, permanent data labels (see figs. 51 & 52 and **6. Label design and production**, fig. 9) can then be added and the distribution of duplicates recorded on the labels.

REFERENCES (those listed under, *List of selected families of flowering plants* and *Tree-climbing techniques* are not repeated).

Anon. (British Museum) (1957)
Allard (1951)
Archer (1945)
Balgooy, van (1987)
Botha & Coetzee (1976)
Camp (1946)
Cooper-Driver & Balik (1979)
Coradin & Ginnasi (1980)
Cremers & Hoff (1990)
Davis (1961)
DeWolf: 8–9 (1968)

Fosberg & Sachet: 29–50 (1965)
Fuller & Barbe (1981)
Gleason & Smith (1930)
Guillarmod (1976)
Hallé (1961)
Hicks (1978)
Hollis et al. (1977)
Jain & Rao: 22–45 (1977)
Leuwenberger (1982)
Lot (1986)
MacDaniels (1930)
Madalski (1958)
Matthew (1981)
Mori et al.: 5–26 (1989)
Radford et al.: 387–398 (1974)
Raynal-Roques (1980)
Sánchez Mejorada (1986)
Savile (1973)
Schnell (1960)
Skvortsov: 52–91 (1977)
Smith (1971)
Steenis, van: 52–53 (1950)
— (1977)
Stern (1965)
Taylor (1977)
— (1988)
ter Welle (1989)
Womersley (1976)
— (1981)

31. COLLECTING MATERIALS FOR ANCILLARY DISCIPLINES

(see also **21. Removal of samples from herbarium material**)

Collecting for ancillary disciplines, e.g. palynology, anatomy, cytology etc. involves special techniques. It is sensible first to contact any specialist for whom you intend to collect material to determine any preference for the type of material, quantity and its method of preservation. If practicable, collect living material or seeds that can be propagated in the recipient institution and then the samples prepared under laboratory conditions. It is important to collect a good voucher specimen so that the identity of the sample can be verified and, if necessary, re-checked by later workers.

Any import restrictions and phyto-sanitary requirements must be observed.

Materials and methods appropriate for the various disciplines

EXAMPLES OF MATERIAL REQUIRED	METHOD/MEDIUM
MORPHOLOGY	
In addition to dry herbarium specimens: – Flowers where the structure would be distorted or obscured by pressing, e.g. Asclepiadaceae, Lentibulariaceae, Orchidaceae. – soft fruits. – succulent taxa such as Cactaceae.	Wet collections made in a fixative such as Kew Mixture or 50–70% alcohol (ethanol, or IMS) and 30–50% water.
PALYNOLOGY	
Mature buds or unopened flowers; check for pollen in the field.	Dry together with herbarium material in folds of paper.
ANATOMY	
Various organs, e.g. fruit, leaf, axils of leaf nerves (for domatia), stem, root.	Fix in 70% ethanol or Carnoy's fluid. AA can be used except for delicate/soft material such as Orchids which need formalin (FAA). Kew mixture can be used but is not ideal.
Very delicate material such as ovules.	Fix in FPA.
Bark*, wood*	Can be air-dried.
Bulbs and fleshy plants such as orchids or cacti.	Simply place in a paper bag; they will usually survive the journey well enough for microscope slides to be prepared. If possible propagate the material on arrival.

* see **30. Collecting and preserving specimens**, *Bark samples* and *Wood samples*.

Materials and methods appropriate for the various disciplines (continued)

EXAMPLES OF MATERIAL REQUIRED	METHOD/MEDIUM
CYTOLOGY: **Karyology**	
Very young sporangia or flower buds (for meiotic studies). Root-tips (for mitotic study) need elaborate pre-treatment before fixation and it is inadvisable to attempt this in the field; better to collect material for propagation.	Fix a range of bud sizes, including smallest to c. ½ full size, in freshly-mixed Carnoy's fluid or 6:3:1 mixture. Split large buds longitudinally to accelerate penetration.
CYTOLOGY: **extraction of DNA**	
Young leaves, but not those that appear too succulent (i.e. contain a lot of polysaccharide) or are too fibrous.	Fresh leaves or leaves dried rapidly. The leaves (1–5 gms.) should be placed in air-tight bottles and the remaining volume completely filled with anhydrous silica gel before sealing with lid and tape. DNA is adversely affected by preservation in alcohols.
PHYTOCHEMISTRY	
Various organs, e.g. leaf, flower, root (including bark), bark, fruit. So-called secondary chemicals can vary seasonally, diurnally and due to local growing and meteorological conditions, particularly in vegetative tissues. For this reason seeds, which have far less variation, are the favoured organ for initial studies.	Air-dry in a draught out of direct sunlight, or preserve in alcohol and ensure that the alcohol is returned to the laboratory with the specimen. (The choice may vary according to which compound is under investigation.) The alcohol method may be obligatory in areas of high humidity. Avoid transporting material damp, which could encourage fungal growth. Cloth or paper containers may be preferable to plastic.

COLLECTING IN LIQUID PRESERVATIVES

Handling chemical preservatives

Many substances combined in preserving and fixing fluids are hazardous to health. Spirits are volatile and flammable, and formalin (formaldehyde in solution) is toxic and possibly carcinogenic, and should be avoided whenever possible. In some countries there may be laws governing the use of hazardous chemicals (e.g. 'COSHH' regulations in the U.K.).

In laboratory conditions substances such as methylated spirits, chloroform and formalin *must* be handled in a fume-cupboard and protective clothing (at least eye protection and surgical gloves) should be worn.

In the field, gloves and goggles should be used and breathing fumes must be avoided. It is possible to purchase small phials of formalin which can be dropped into the jar of spirit mixture and then opened remotely. It is important to remember that most airlines do not allow alcohol over 70% strength to be carried, although small quantities may be allowed in the cabin. To satisfy the law of countries with religious prohibitions against alcohol it may be necessary to make sure the spirit has been denatured (i.e. made poisonous and unpalatable) so that it cannot be drunk.

Chemical fixatives and storage media

MEDIUM (and alternative names)	**FORMULA** and/or **FORMULATION**	**PROPERTIES**	**NOTES**
INDUSTRIAL METHYLATED SPIRIT (IMS), denatured alcohol	CH_3CH_2OH ethanol +CH_3OH (methanol)	Flammable over 70% conc., poisonous, volatile Shelf-life indefinite Store in airtight bottles	*Alternatives*: Ethanol (ethyl alcohol), isopropyl alcohol (rubbing alcohol), gin, whisky, vodka, rum and other spirits!
AA	1:18 mixture of glacial acetic acid and 70% ethanol	Flammable, volatile, poisonous	Generally preferable to FAA as it lacks formalin which is a health hazard. After fixation for 2–3 days transfer to 70% ethanol for storage
FAA (Formalin-acetic acid-alcohol)	1:1:18 mixture of 40% formaldehyde, glacial acetic acid and 70% ethanol	Corrosive, flammable, volatile, poisonous Shelf-life indefinite Store in bottles with plastic lids or stoppers	*Alternative*: FPA After fixation in FAA for 2–3 days, or up to 7 days with large volumes of bulky material, material can be transferred to 70% ethanol for storage

Chemical fixatives and storage media (continued)

MEDIUM (and alternative names)	**FORMULA** and/or **FORMULATION**	**PROPERTIES**	**NOTES**
FPA (Formalin-propionic acid-alcohol)	1:1:18 mixture of 40% formaldehyde, glacial propionic acid and 70% ethanol	As FAA Hardens material less than FAA	*Alternative*: FAA The traditional fixative for histological (tissue structure) rather than cytological (cell structure) use Propionic acid is a laboratory chemical and may be harder to obtain than acetic acid and is more expensive
ACETIC ACID, glacial (Ethanoic acid)	CH_3COOH	Corrosive, poisonous Shelf-life indefinite Store in glass or plastic bottles with plastic lids or stoppers Freezes at 16°C	Used in the formulation of AA, FAA, Carnoy's fluid and 6:3:1 mixture
CARNOY'S FLUID	3:1 mixture of absolute alcohol (96%) and glacial acetic acid	Not to be mixed until required. Refrigeration is required for long term storage, but material can be kept at room temperature for 2–3 weeks	*Alternative*: 6:3:1 mixture for cytology Fixative may be replaced with 70% ethanol after fixing for 24–48 hours
6:3:1 MIXTURE	6:3:1 mixture of chloroform, 95% ethanol and glacial acetic acid	The glacial acetic acid should not be added until material is ready to be fixed Volatile, flammable	*Alternative*: Carnoy's fluid Replace fixative with 70% ethanol after fixing for 24–48 hours

Chemical fixatives and storage media (continued)

MEDIUM (and alternative names)	**FORMULA** and/or **FORMULATION**	**PROPERTIES**	**NOTES**
KEW MIXTURE	10:1:1:8 mixture of IMS, formalin, glycerol and water	Poisonous, volatile	*Alternatives*: FAA, FPA or 50–70% alcohol can be used but does not fix material as well
COPENHAGEN MIXTURE	10:1:8 mixture of IMS, glycerol and water	volatile	A storage mixture. The glycerol helps keep the material pliable

Collecting portions of plants into liquid preservatives

(see also **13. Ancillary collections** and **30. Collecting and preserving specimens**)

Equipment

- Leak-proof plastic bottles. Before going in the field – fill full of water, fasten lid and squeeze to test for leakage – take a sealant (e.g. paraffin wax) if necessary. A full range of sizes, including a wide-mouthed 'base-camp' bottle will be needed. Also small glass phials for flower buds etc.
- Bulk containers of spirit and supplies of other chemicals.
- Small, pre-punctured polythene bags or muslin for wrapping around individual specimens.
- Cotton or linen thread or fine cord to fasten polythene or muslin wrappings.
- Number tags and pencils or pens with spirit-proof ink.
- Scissors, secateurs and knives for cutting and preparing the material.
- Goggles and surgical gloves (important if formalin is to be used).
- Boxes and packing material.

Method

Specimens for morphological study are best if they have been *fixed* when collected. The purpose of a *fixative* is to kill the tissues quickly and to minimize distortion. Good fixation is essential for fleshy flowers such as orchids. For ideal results a recommended fixative is Kew Mixture: 10:1:1:8 mixture of Industrial

Methylated Spirit (IMS), formalin, glycerol and water. In the absence of a good fixative, 50–70% alcohol (ethanol or IMS) can be used as a substitute, but this does not produce results as good as with Kew Mixture. For permanent storage the material is later transferred to Copenhagen Mixture.

It is important that specimens for anatomical and cytological study are appropriately and adequately fixed in one of the recommended media (see table above) before transference to a storage medium.

In the field, liquid weight can be saved in the case of more robust items by using lightly pre-punctured bags or by wrapping in muslin to decrease damage in transit. Place these, and any (protected) small bottles and tubes in a *wide*-mouthed bottle containing a small quantity of the preservative which can be topped up when convenient. Large fruits can be kept in well sealed polythene bags with sufficient preservative to keep them thoroughly wet. In all cases labels bearing the collection number should be included with each collection. If field time is *very* limited, keep the samples with the rest of the plant until base camp is reached and then check and properly number the contents and containers.

It is not advisable to handle wet material (especially flowers), so if multiple sets of spirit material are to be distributed, separation of the material is best done in the field, ideally at the time of collection. Note that any agreement with the host country must be honoured, and if they request that a proportion of the collections remains in the country this includes spirit material. Nevertheless, wherever possible separately stored material is best sorted and checked for consistency and extra labels and cross-references added at the collector's institution. See Lawrence 1951; Radford et al. 1974; Tomlinson in Fosberg & Sachet 1965; Womersley 1981.

REFERENCES

Lawrence: 253–256 (1951)
Morawetz (1989)
Radford et al.: 388–389 (1974)
ter Welle (1989)
Tomlinson in Fosberg & Sachet: 117–119 (1965)
Womersley (1957)
— :35–40 (1981)

32. THE COLLECTION OF PTERIDOPHYTES

The pteridophytes consist of a large group, the ferns, and a smaller group, the fern allies. Although the two groups are not closely related, they are nevertheless similar in general physical characteristics, tend to grow in the same habitats, and are grouped together in herbarium arrangements. It is thus useful to treat them together for collecting purposes.

The essential factor in collecting pteridophytes is to make sure the specimens are fertile, that is, bearing spores. The spores are borne in sporangia which differ in appearance and position on the plant in the various groups of pteridophytes. The sporangia are borne on bare branches in *Psilotum*, a fern ally, or in the axils of 'leaves' in *Tmesipteris*, *Isoetes* and some *Lycopodium* species (all are fern allies), or aggregated into cones or spikes in *Selaginella*, many *Lycopodium* species, *Phylloglossum*, *Equisetum* (all fern allies) and in Ophioglossaceae, a primitive family of ferns. In the majority of ferns the sporangia are borne on the margin or under surface of the frond (the leaf-like photosynthetic organ of the plant) where they are sometimes protected by indusia, which are specialized outgrowths of the frond tissue. There is no point in collecting pteridophyte material which is not fertile as it is usually extremely difficult or impossible to identify, even by experts.

Ideally the fertile material will be mature, that is, with ripe spores, and these can be detected in most ferns in the field with a little practice. A ×10 hand-lens should always be carried in the field and with the aid of it details of the dark brown sporangia full of minute ripe spores can be seen. A few families, particularly Osmundaceae, Hymenophyllaceae and Grammitidaceae have green, photosynthetic spores and these too can be identified in the field with a little practice. Although, in general, with these families it is not essential to have mature fronds, they should, of course, be fertile. Similarly, with practice the mature sporangia of the fern allies will also be recognizable. Such mature sporangia will usually shed their spores during the drying process and this is an indication that the specimen is in good condition, providing the maximum amount of information. Some of the spores shed during the drying process should be packeted and included in a capsule as they may be used for SEM studies and are more likely to be mature and thus properly developed than those remaining within the sporangia. The techniques of preserving, pressing and drying pteridophyte specimens are the same as for higher plants (see **30. Collecting and preserving specimens**). Pteridophyte fronds are usually two-dimensional in plan and are therefore easier to press than many higher plants.

Although it is usually possible to identify a single fertile frond of most pteridophytes, in practice it is much more useful to collect portions of the rhizomes as well, because their habit, whether erect or creeping, and their covering of scales and hairs are important taxonomic characters in many groups. Just the top half of a fertile fern frond on its own should never be collected.

The size range within ferns is as great or greater than that in any other group of plants. The fronds of some species of Hymenophyllaceae (filmy ferns) are less than 1 cm. long, while those of some Cyatheaceae (tree-ferns) and *Angiopteris* are up to 7 m. long. For the purposes of collection it is convenient to divide pteridophytes into 3 groups according to the size of the plant; rather different techniques are required for each group.

Small ferns, where several to many fronds or plants will be needed to fill one herbarium sheet adequately. These will have either long-creeping rhizomes, in which case they may often form loose to dense mats, or they will have short more or less erect rhizomes. With the former it is necessary to check carefully that only one species is included in a specimen; some mat-forming ferns such as Hymenophyllaceae will grow as a mixture of species with rhizomes intertwined. With the latter it is necessary to check carefully that all plants collected under the one number belong to the same species. Examination in the field with a ×10 hand-lens will sort out the most obvious problems, but for Hymenophyllaceae in particular it is wise to check material under a higher-power binocular microscope in the herbarium to ensure that only one species is included under each number. Ideally, these small ferns should be placed in individual plastic bags in a larger plastic bag in the field to avoid contamination of collections. If small plastic bags are not available, toilet paper is a reasonable substitute to keep them separate. It is useful to dry these collections in individual newspaper packets to ensure that the small fronds do not drop out of the press. It is possible to collect wilted material of Hymenophyllaceae and soak it in water for a few hours, then press and dry in the usual way.

Medium size ferns, where a whole frond will fit on one or two herbarium sheets, are the easiest to deal with, as a plant will often provide enough fronds to make duplicates and there is no doubt that all specimens of that number refer to the same taxon. Where the rhizome is long-creeping, enough should be taken to indicate how far apart the fronds are spaced. If the rhizome is erect the whole plant may be taken as a specimen and cut up in such a way that each frond is attached to a section of rhizome. If space permits it is better to cut long fronds into two at the time of pressing rather than fold them over, as this usually leads to some bruising and folding of the lamina (or blade of the frond), where it is pressed against the stipe (stalk of the frond); but if this is done both portions should be tagged with the collecting number. If the frond is too wide to fit on the herbarium sheet the pinnae (primary divisions of the lamina) on one side may be trimmed almost to the rachis (lamina midrib), leaving just enough to indicate their position in relation to each other and the remaining pinnae.

Large ferns, where it is impractical to collect and/or preserve complete fronds, are usually poorly represented in herbaria, although it is relatively simple to make good specimens of such large fronds, which can be mounted on only three herbarium sheets. In general the frond apex or a couple of pinnae are taken but this is rarely adequate for identification of critical groups. If the rhizome is creeping, the distance apart of the stipes should be recorded in the field if it is too great to fit into the press. It is useful to get a portion of the rhizome from near its apex to ensure that its scales or hairs are in good condition, as they are often broken when old. If the rhizome is erect, the base of the stipe should be cut away from the rhizome to ensure that the scales are included, and it is also useful to take some of the very young fronds with undamaged scales. Enough of the base of the lamina should be collected to make a standard-sized herbarium specimen, cutting away the pinnae on one side as described above if necessary. The central portion of the lamina should be taken, and also the apex. Every portion of the plant collected should be labelled with the collection number. If time permits, a photograph of the complete frond can be taken prior to cutting up, and if space permits, the complete frond can be prepared as a series of herbarium specimens, the sections being numbered in sequence so that the frond can be reassembled, if needed, when dried and mounted on herbarium sheets. If duplicates are being made, each should be prepared from a single frond from the same plant. If the plant is a tree-fern and is common, it is useful to cut it down and make cross sections of the trunk c. 1 cm. thick which can be dried with care and mounted on the herbarium sheet. Larger sections of the trunk may also be dried and included in carpological boxes. The information needed on these large ferns, apart from the usual habit and habitat notes, includes:

- The length and width of the frond and the length of the stipe and, if a tree-fern is collected, the height and diameter of the trunk.
- The presence or absence of aerial roots.
- The number of fronds.
- Whether the fronds are produced sequentially (one after another) or synchronously (in whorls and all at approximately the same stage of development, as in most tree-ferns).
- How many there are in a complete whorl.
- For tree-ferns: the angle at which the fronds are held to the trunk.
- If they are shed from the trunk when dead or if they are retained as a skirt.
- How far apart they are spaced on the trunk.

This may appear to be a tedious procedure in the field, but it adds immeasurably to the usefulness of the herbarium specimen.

Fern spores can be collected for germination (see Roos & Verduyn 1989).

REFERENCES

Arreguín-Sánchez (1986)
Henty (1976)
Holttum (1957)
Mickel (1979)
Roos & Verduyn (1989)
Stolze (1973)

33. THE COLLECTION AND CURATION OF BRYOPHYTES*

The bryophytes include the mosses (Musci), the liverworts (Hepaticae) and the hornworts (Anthocerotae). These classes share a number of features and are unlikely to be confused with any other group of plants, except perhaps some small Pteridophytes. They are generally small, non-vascular land plants with a unique life cycle involving two distinct stages – a long-lived, green phase (gametophyte) which reproduces sexually and a short-lived, dependent phase (sporophyte) which is not green at maturity and reproduces by means of spores.

Collection

Bryophytes are among the easiest of plants to collect and preserve. Specimens are best collected directly into paper packets on which a reference number and details of locality and habitat can be written. Polythene bags should not be used, except sometimes as a very temporary measure, since high humidity inside the bags will not only encourage the development of moulds but can promote the growth of abnormal, etiolated shoots.

The quantity of material to be collected will obviously depend on the amount present, but should also take account of the need for conservation of both the species and the community of which it is a part. Where possible, sufficient should be gathered to permit the preparation of several duplicates which can then be exchanged with other institutions. Sporophytes should be included whenever possible; they are not always readily detectable, especially in liverworts, and examination with a lens may be necessary to confirm their presence. Small species are easily overlooked and should be carefully sought, e.g. on tree trunks, branches, rocks, soil and, occasionally, evergreen leaves. (Note that the more interesting species may occur in habitats that look relatively unpromising, rather than in those dominated by moss growth).

Specimens should be dried as quickly as possible, using artificial heat only if necessary, and always at a fairly low temperature. To facilitate this, the material should be spread out within the collecting packets and the latter partly opened and laid out in a well-ventilated place. Leaves supporting liverworts should be dried under light pressure (i.e. as for vascular plants). Delicate plants such as some leafy liverworts, and fleshy plants such as some thalloid liverworts, must not be allowed to overheat during the drying process or the resultant 'cooking' will distort cell structures, preventing reconstitution of the specimen on re-wetting. In the case of earth-growing species, as much surplus soil as possible should be removed before drying (but leaving the rhizoidal mat which may contain rhizoidal gemmae or tubers).

* Contributed by Alan Eddy, of the Natural History Museum, London, and Alan Harrington, formerly of the same.

Curation

Dry bryophyte specimens are relatively immune to insect or fungal attack. It is therefore unnecessary to treat them with poisons; indeed, they can be ruined by some such treatments. In tropical herbaria, where there is a potential risk of fungal infection, air-conditioning and de-humidification may be essential.

When thoroughly dry, each specimen is placed in a new packet, then filed or 'laid-in' along with all of the relevant data, the latter typed or legibly written either on the packet itself or on an attached label. In most large herbaria, packets are mounted on standard herbarium sheets which are arranged in systematic or sometimes alphabetical order. An alternative procedure is to arrange the packets in filing drawers or boxes. If the latter method is used, it is strongly advisable to use a standard packet size, since small packets can become hidden between or within the larger ones.

Special problems arise in the case of mixed gatherings, when some system of cross-referencing may prove necessary.

REFERENCES

Buck & Thiers (1989)
O'Shea (1989)

34. THE COLLECTION AND CURATION OF ALGAE*

The Algae are a highly diverse group of plants found in damp terrestrial habitats, in freshwater and in the sea; a few taxa are adapted to survive in arid conditions. This account deals with the collection and preservation of seaweeds (benthic, marine macro-algae) and stoneworts (charophytes) only.

Seaweeds grow principally in the intertidal and subtidal zones of rocky shores, but also occur in salt marshes and mangrove swamps. Three major groups are recognized, the Chlorophyta (green algae), the Phaeophyta (brown algae) and the Rhodophyta (red algae). These vary considerably in form and size and this variation can cause problems in preservation and curation. For a detailed account of the morphology and ecology of the seaweeds consult a suitable textbook, for example: Dawson, 1956; Chapman and Chapman, 1973; Dawes, 1981; Lüning, 1990.

The charophytes (Characeae), the only freshwater algae treated here, are complex, medium to large Chlorophyta with whorled branching and habit similar to *Equisetum*. They grow submerged in fresh and brackish waters, and are particularly common in disturbed and newly created habitats.

Collection of seaweeds

Working on the sea-shore requires care and attention. To begin with, the Tide Tables should be consulted so that collecting can be arranged when low water occurs at the optimum time of the day; it is safer to work during falling tides than rising tides. The weather forecast should be checked as poor weather conditions will, on occasion, impede successful sea-shore and off-shore activities, and may also prove dangerous. Adequate clothing should be worn; in colder waters thigh boots (waders) will give protection in deeper pools, while in warmer conditions one need only wear plastic sandals (or similar light shoes) to provide protection against barnacle cuts and injury from more dangerous underwater objects. Shallow sub-tidal levels can be investigated by snorkelling; greater depths are best investigated by SCUBA diving or, less effectively, by dredging. Strict attention to health and safety regulations is essential during diving and boat-handling activities, which should **only** be attempted by fully trained personnel.

The equipment required for collecting specimens includes:

- Plastic buckets, bottles and bags.

- Plastic tubes and small 'snap-top' plastic bags for separating small or special specimens.

* Contributed by Jenny A. Moore, Curator of Algae, the Natural History Museum, London.

- Knife or metal scraper, needed for removing material from rocks.

- Hammer and chisel for collecting crustose species.

- x10 hand-lens is helpful for separating different taxa in the field.

When collecting, attention should be paid to the less obvious species that grow on, or under, the larger, conspicuous plants. Small, delicate forms may be found as epiphytes on the fronds, stipes and holdfasts of large thalli, whilst crustose taxa grow both as epiphytes on other algae and as lithophytes on hard substrata. Some species grow only submerged in rock-pools; filamentous forms may bind sand and silt from which they must be washed gently.

Data to accompany seaweed specimens should include:

- A general description of the habitat including position on the shore (e.g. H.W.S.T = High Water Spring Tides), substrate type and degree of exposure to wave action.

- Associated plants and animals.

- The name of the 'host' if the alga is attached to another plant or animal.

- The depth if the specimen is collected in the subtidal zone.

If seaweeds are to be examined within 24 hours of collection, they are best kept in sea-water and examined fresh, since characters such as cell and chloroplast shape are often lost on drying. Nevertheless, the conventional method of pressing and mounting specimens on herbarium sheets is used with seaweeds, although the technique is specialized (see below). Some species of seaweed start to decay immediately on collection and will cause deterioration of other specimens; pigment leaching into the water may indicate this and the sample should be preserved at once.

Collection of charophytes

Charophytes can be collected directly by hand, wading in shallow water or by snorkel or SCUBA diving in deeper water (see notes above on seaweeds, also Lot 1986, figs 1 & 2). A grapnel or dredge can be used for direct and remote sampling by boat in deep water. The most suitable sampler is a three or four-pronged grapnel, with the long prongs bent backwards. Other useful samplers include a right-angled digging fork, or a sharp-edged scoop attached to an extending handle. In relatively clear and shallow water these samplers can be directed to specific areas of the bottom by viewing through a glass-bottomed bucket or a diving mask.

Data to accompany charophytes should include:

– General description of habitat including size and type of water body (e.g. small temporary pool, large lake or sluggish stream) and substrate type (pH, if known).

– Associated plants.

– Depth range of collected species (if visible).

After collection, specimens should be rinsed and excess water removed before storing in polythene bags together with the collection data written on suitable, good quality paper. Specimens will keep fresh for a few days if the bags are stored in a cool, dark place (or preferably a refrigerator). However, if they are needed for further study, they should be preserved as soon after collection as possible.

Preservation

Preparation of herbarium specimens

The following equipment is required:

– A shallow dish or tray (photographic developing trays are useful).

– Heavy-duty cartridge or similar paper which is strong enough to maintain its rigidity when wet.

– Strong wire mesh or rigid plastic sheet.

– Fine muslin, Nylon Gossamer Fabric*, or nylon stockings/tights.

– Mounted needle, forceps, small soft paint brush and pencil.

– A plant press, with drying paper and corrugates.

The method for preparing dried pressed specimens of **seaweeds** is as follows:

1. Annotate mounting paper with data in pencil, place on wire mesh and immerse in the tray, two-thirds full of sea-water.

2. Float specimen on to mounting paper.

3. Tease out thallus to display branching pattern or form of the plant, using forceps, mounted needle and brush.

4. Lift one end of wire mesh, supporting the paper and specimen, allowing water to drain away.

* Supplied by Picreator Enterprises Ltd., 44 Park View Gardens, Hendon, London NW4 2PN, U.K.

5. Make minor adjustments to mounted specimen, if necessary.

6. Remove paper with specimen from supporting wire mesh and place on absorbent drying paper.

7. Most plants will adhere to the mounting paper by means of their own natural mucilage. It is, therefore, important that the specimen is covered with muslin or nylon fabric to prevent it sticking to the upper drying paper.

8. Place drying paper on top and continue the process, placing corrugates at intervals, before putting in a plant press.

9. It is important to dry the specimens quickly, but without excessive heat or they may become brittle.

As it may be necessary to remove samples of small specimens from the mounting paper for microscopic examination, it is helpful if some branches are left clumped together for ease of removal, rather than spread out finely. Unattached parts of specimens can be secured using the appropriate adhesive or gummed fabric tape. Specimens that remain completely detached from the mounting paper should be put in paper capsules. Pressed specimens, capsules and labels can then be mounted on the herbarium sheets in the usual way.

Charophytes are prepared for the herbarium by similar 'floating out' techniques, except that fresh (tap) water should be used instead of sea-water. They benefit from washing with a jet of water from a pipette to remove snails and detritus. Some charophytes are lime-encrusted, so only light pressure should be applied in the press. Not all adhere naturally to mounting paper, so they need to be strapped or glued in place. As charophytes are often delicate or brittle they should be protected when mounted on herbarium sheets. Small specimens should be put in paper capsules and larger specimens should be covered with protective film, such as '*Melinex*' or '*Mylar*' (see **5. Materials**, p. 44), or thin paper. This should be attached to the herbarium sheet on two sides only (bottom and left-hand side of specimen) to form an easily opened covering.

Articulated **Corallinaceae** (marine algae: Rhodophyta) that contain high concentrations of calcium carbonate are so fragile that they will be damaged by pressing and should be air-dried. Other bulky **crustose species** attached to rock or hard substrates should be similarly treated. The surfaces should first be gently brushed to remove debris and small animals. Air-dried specimens are best stored in boxes or packets that give some protection.

Generally, seaweeds are not prone to damage by insect pests and decontamination for herbarium beetle etc. is unnecessary. However, large, slow-drying specimens are prone to fungal attack, particularly in humid weather. This can be arrested by brushing the specimen with 4% formalin (see below). Similarly, air-dried articulated corallines and crustose algae may need to be so protected whilst they are drying.

Liquid preservation of algae

Any algal material that is to be stored for periods longer than 24 hours requires fixation and preservation by 4% neutralized formalin (1 part 40% formaldehyde diluted with 9 parts of water (sea-water for seaweeds)). This is a potentially **dangerous** chemical and care is needed when using it. Prolonged handling should be avoided. Neutralization is necessary to prevent the formation of formic acid; if allowed to build up in the solution the acid will eventually damage the specimens. The formaldehyde solution should be neutralized using borax or calcium carbonate; a little glycerol may be added to prevent specimens drying out. Specimens kept in this way should be stored in water-tight, glass jars in a dark, well-ventilated place.

Similarly, charophytes can be preserved in 4% formalin, or in 70% alcohol (3 parts of 95% ethyl alcohol (IMS): 1 part water). Dried specimens can be more difficult to identify than those preserved in liquid. FAA (see **31. Collecting materials for ancillary disciplines**, p. 226) is not recommended as the acetic acid removes the lime encrustation, which is a helpful diagnostic feature in some taxa. The liquid-preserved collection should be stored as for seaweeds but, since those preserved in alcohol are prone to evaporation, regular inspection and topping-up will be necessary.

REFERENCES

Chapman & Chapman (1973)
Dawes (1981)
Dawson (1956)
Lot, figs 1 & 2 (1986)
Lüning (1990)

35. THE COLLECTION AND PRESERVATION OF FUNGI (INCLUDING LICHENS)

Collection and preservation of fungi is a specialist undertaking, and scientifically useful material can only be obtained using techniques different from those employed by the general plant collector. An outline of these techniques is provided below. To assist those unfamiliar with fungi and their special terminology, the principal types of larger fungi are briefly described and illustrated at the end of this chapter together with a glossary of terms.

COLLECTING FUNGI

Equipment and method

Collections, particularly of fleshy fungi, should be placed in a flat-bottomed basket or open box to minimize damage to specimens. They should be kept distinct in the field, preferably by wrapping in waxed paper; if this is not available small boxes could be substituted; both polythene and newspaper should be avoided. Smaller or more delicate species and microfungi should be placed in individual containers (plastic food containers, e.g. yoghourt pots, are ideal). It is important to avoid excessive handling of specimens. Some surface features are delicate or ephemeral and easily lost.

What to collect

It is important that only specimens in good condition should be collected. Abundant gatherings of each species should be made wherever possible in order to show all stages of development. For agarics it is essential that the entire fructification is collected including the base of the stipe and any volva which may be present.

Parasitic species must be collected with sufficient material of the host plant to enable this to be determined as well as the fungus. In some circumstances, e.g. in under-collected areas, it may be desirable to collect a good voucher specimen of the host.

Numbering the collections — see **30. Collecting and preserving specimens**, *Numbering collections.*

Recording data

See **30. Collecting and preserving specimens**, *Recording data.*

In the case of fungi, details of the habitat should be given including: habitat type, associated plants, type of soil, substrate or host plant as appropriate. Careful descriptive notes should be made when the material is fresh. Specimens should be handled with care to avoid loss of delicate surface structures. The notes must include:

- *Overall description.* Give descriptive notes on the general appearance of the fruit-bodies; sketches can be very useful. Such notes must include the type of fertile surface, such as smooth, gills, teeth, pores. Details of the gill attachment to the stem in agarics is important (see Fig. 60B).
- *Size range.* Length and width of all structures of mature specimens.
- *Colour range.* Details of all parts of the fruit-body (young and old); colour changes on bruising, drying or freshly broken flesh; colour of any latex which may be present. It is advisable to use a published colour chart.
- *Surface features.* Note whether the surfaces are smooth, scaly, hairy, sticky or viscid, striate, or have any other distinguishing features, Be sure to record the presence or absence of any veil structures, such as ring and/or volva on the stem.
- *Texture.* Note the consistency of the flesh, e.g. thick, thin, pliable or brittle, fibrous, fleshy or woody.
- *Odour and taste.* Many species have a highly characteristic smell which can only be detected in fresh specimens. Taste is also very important, e.g. peppery, bitter. Care must be taken never to swallow the flesh; **remember some fungi are deadly poisonous**, and tasting should **only** be attempted by a collector with some knowledge of the local fungi.

A water-colour sketch or, alternatively, a colour photograph of the fresh specimen is also desirable, particularly for fleshy fungi. This should show the form of the fruit-body and, in agarics, a vertical section to show the type of gill attachment. It should be emphasized that the photograph/painting is valuable but does not replace the written description.

Because of the time needed to obtain the necessary information for each collection only as many species as can be fully documented in the available time should be collected.

Spore deposits (prints)

A spore deposit (spore print) from at least one specimen of each collection is desirable for larger fungi. Spore deposits usually take at least two hours for the spores to be deposited in sufficient quantity. They may be prepared in the following ways:

- **Basidiomycetes**: Place the fruit-body with the hymenial surface downwards onto a sheet of paper or glass slide and leave overnight (in a cool place). The specimens should be covered or, preferably, placed in a tin or under a glass jar to prevent desiccation, with a drop of water placed on the pileus.

For *agarics* it is best to remove the stipe before taking a spore deposit. For species of agarics with coloured gills, spore deposits should be taken on white paper, but deposits of species with white gills are better obtained on dark coloured paper. In the latter instance, it is often helpful to draw a ring around the deposit, particularly if it is scanty.

Some *polypores* and *resupinate* fungi have very short periods of sporulation and the spores of some of these have yet to be described. For these, spore deposits are particularly valuable. If too dry, spore deposits from material of these fungi may often be successfully obtained following rehydration of the collection by placing it in a closed container with wet cotton wool.

- **Ascomycetes**, particularly larger *discomycetes:* as above but place a glass slide above the hymenial surface as close to the fungus as possible.

Spore deposits are air-dried and kept with the specimen (see **17. Curation of special groups**, *Fungi and lichens)*. Spore prints and specimens should both be clearly numbered so that there is no risk of them being mixed up.

Drying fungi

It is important that specimens are dried as rapidly as possible after collection as this helps to retain form and colour and prevents deterioration of the material. Small agarics may be dried whole, but larger ones should be halved to facilitate drying. Polypores are generally easier to dry than agarics, but may need to be cut into portions.

Successful and rapid drying ideally requires a constant flow of warm air over the specimens, for which a temperature of c. 40°C is recommended. Care should be taken that the temperature does not become too high, as the specimens may cook. Too low a temperature will only succeed in hatching insect larvae which will eat the specimens. The specimens may either be taken back to the laboratory and placed in a drying cabinet or dried on a portable field stove at the base camp. Small fungi may sometimes be successfully air-dried in the sun.

Warm air flow can be obtained in the laboratory by using an electric fan heater, or a 60W lamp in a perforated tin, placed in a frame beneath a number of wire shelves. In the field, a similar perforated tin or wire shelves may be suspended over a fire, or an oil lamp or stove. Burners fuelled from gas cylinders also provide a very suitable means of heat. Specimens should be suspended across a frame erected around the burners. See also p. 10, *Drying facilities* and p. 213, *Field stoves.*

In localities with high atmospheric humidity, the thoroughly dried specimens must be stored either in a desiccator over calcium chloride, or sealed in polythene bags. In the latter case, a few granules of silica gel should be added where possible. Clip-top polythene bags are most successful for storing collections in the field, and ideally a number of smaller bags should be stored inside a larger clip-top bag. In all cases it is essential that the specimens are thoroughly dried, and checked in the field at regular intervals. If any dampness is present, specimens will soon be ruined by moulds and bacteria.

Fungi may be freeze-dried for display but the method has no other advantages. The equipment needed is expensive, and freeze-dried specimens are brittle and often bulky.

Collections of rust and other plant pathogenic fungi made on the host plant can be treated in the same way as flowering and other vascular plants, and pressed until dry.

Preservation in spirit mixture

Spirit specimens are occasionally useful to supplement dried material. This is not normally necessary for agarics, and for the polypores only small slices are required. However, all phalloid (Stinkhorns) and clathroid (Cage Fungi) fungi, for which the form of the fruit-body is an important taxonomic character, can be preserved as entire fructifications in spirit. It is desirable to use a fixative for fungi, see **31. Collecting materials for ancillary disciplines**, *Collecting into liquid preservatives*.

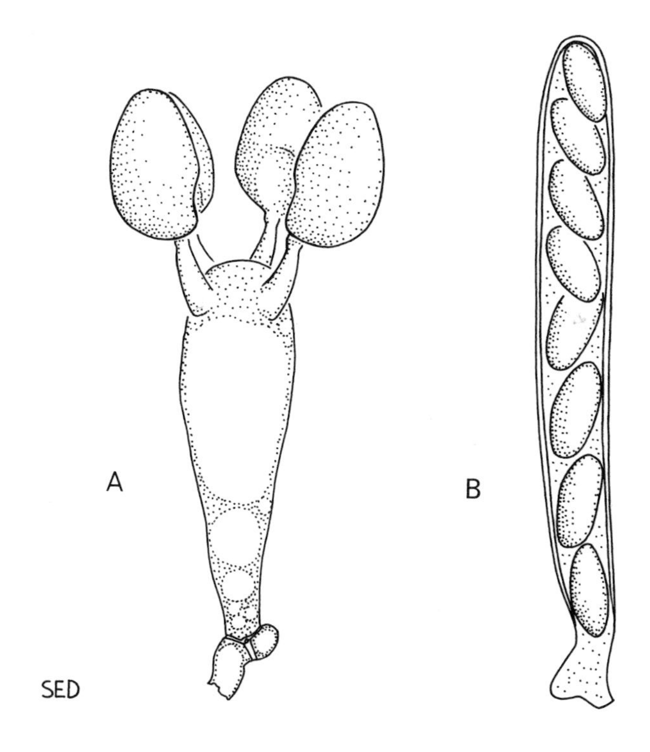

FIG. 59 — **A** Basidiomycetes, a simple basidium bearing spores; **B** Ascomycetes, an ascus containing spores. Both microscopic details.

It must be stressed that preservation in spirit leaches the colour of specimens and is liable to render ineffective subsequent chemical reactions, which may be important tools in identification. Spirit is not recommended for general use, and it is essential that detailed colour notes are made on fresh material which is destined for spirit preservation. Shipment of spirit-preserved collections is facilitated if these are wrapped in impregnated cotton wool and packed in polythene bags rather than being placed in tubes or jars of liquid.

For herbarium curation of fungi and lichens see **17. Curation of special groups.**

PRINCIPAL TYPES OF LARGER FUNGI

Fungi are extremely diverse and very numerous and it is not possible to give more than a basic guide to the classification of those fungi which produce large fruit-bodies. These are classified into two main groups, **Basidiomycotina** ('Basidiomycetes') and **Ascomycotina** ('Ascomycetes').

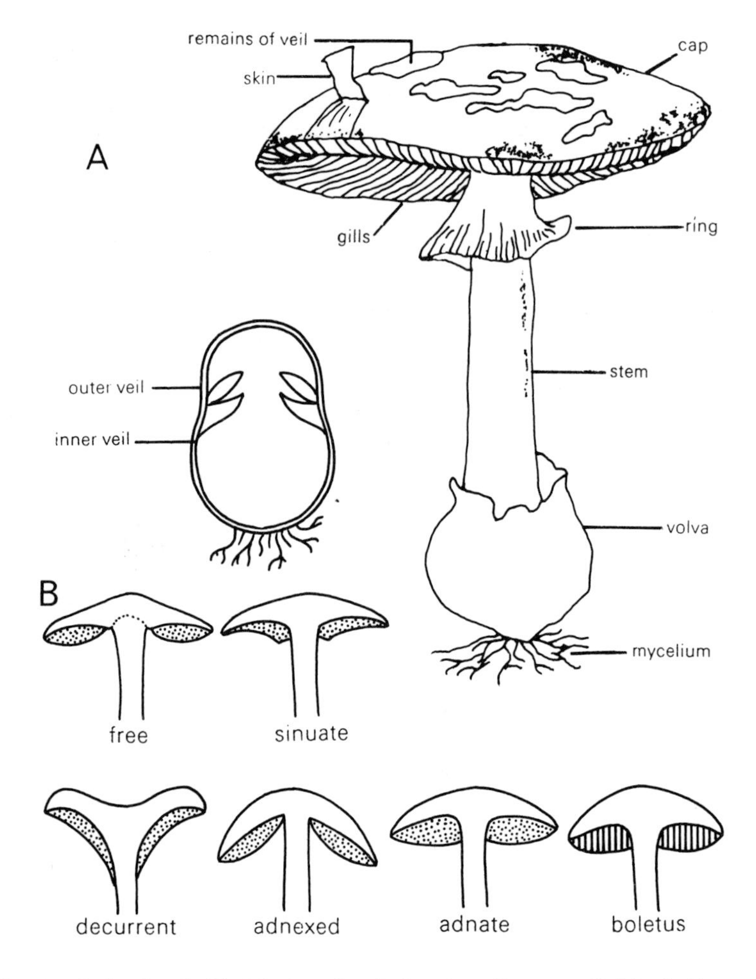

FIG. 60 — Agaricales: **A** typical fruit-body and vertical section of immature fruit-body; **B** types of gill attachment and tubes. The latter are found in *Boletus* and a few other genera.

Basidiomycotina. Spores formed externally on a *basidium* (pl. *basidia*). (Fig. 59A).

1. *Agaricoid Fungi* (Fig. 60) – Mushrooms and toadstools (or agarics), including the boletes; soft-fleshy fruit-bodies, quickly rotting; spores produced under the cap on radiating gills or vertical tubes each opening as a pore.

2. *Poroid Fungi* (Figs. 61 & 62A) – Bracket Fungi (polypores); typically bracket-like, commonly growing on wood, sometimes with a stem; tough fruit-bodies, long-lasting and sometimes perennial; spores produced under the cap, when present, within a layer(s) of vertical tubes.

3. *Stereoid Fungi* (Fig. 62E) – Skin Fungi (*Stereum* and allies); bracket-like or resupinate fruit-bodies, with a smooth fertile surface.

4. *Clavarioid Fungi* (Fig. 62B & C) – Fairy-clubs and Coral Fungi; simple or branched club-shaped fruit-bodies, soft and short-lived but not gelatinous, with a smooth fertile surface, often brightly coloured.

5. *Hydnoid Fungi* (Fig. 62D) – Tooth Fungi (*Hydnum*) and similar species; fertile surface with teeth or spines; fruit-bodies may be fleshy or tough, with or without a stem.

6. *Gasteroid Fungi* (Figs. 62H–M) – Puffballs, Earthstars, Earthballs, Bird's Nest Fungi, Stinkhorns and Cage Fungi. An unnatural assemblage having in common an enclosed spore development and passive spore release.

7. *Jelly Fungi* (Figs. 62F & G) – A mixed group characterized by gelatinous fruit-bodies and distinctive basidia. Three orders are commonly recognized: *Tremellales* (vertically septate basidium), *Dacrymycetales* (tuning-fork basidium); *Auriculariales* (transversely septate basidium).

8. *Rust Fungi* (*Uredinales*) and *Smut Fungi* (*Ustilaginales*). Plant parasites.

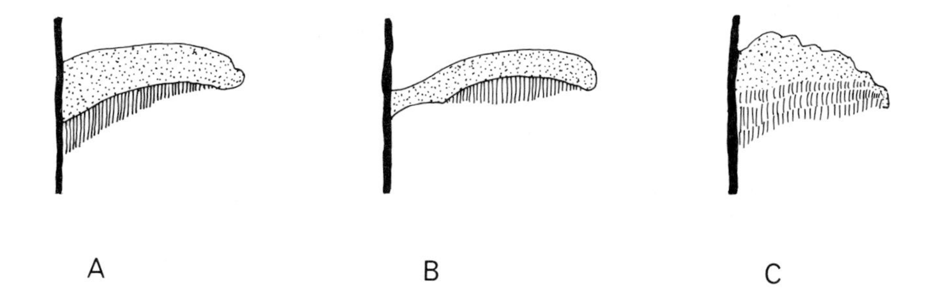

FIG. 61 — Polypore attachment. **A** sessile; **B** stipitate; **C** sessile perennial with stratified tubes.

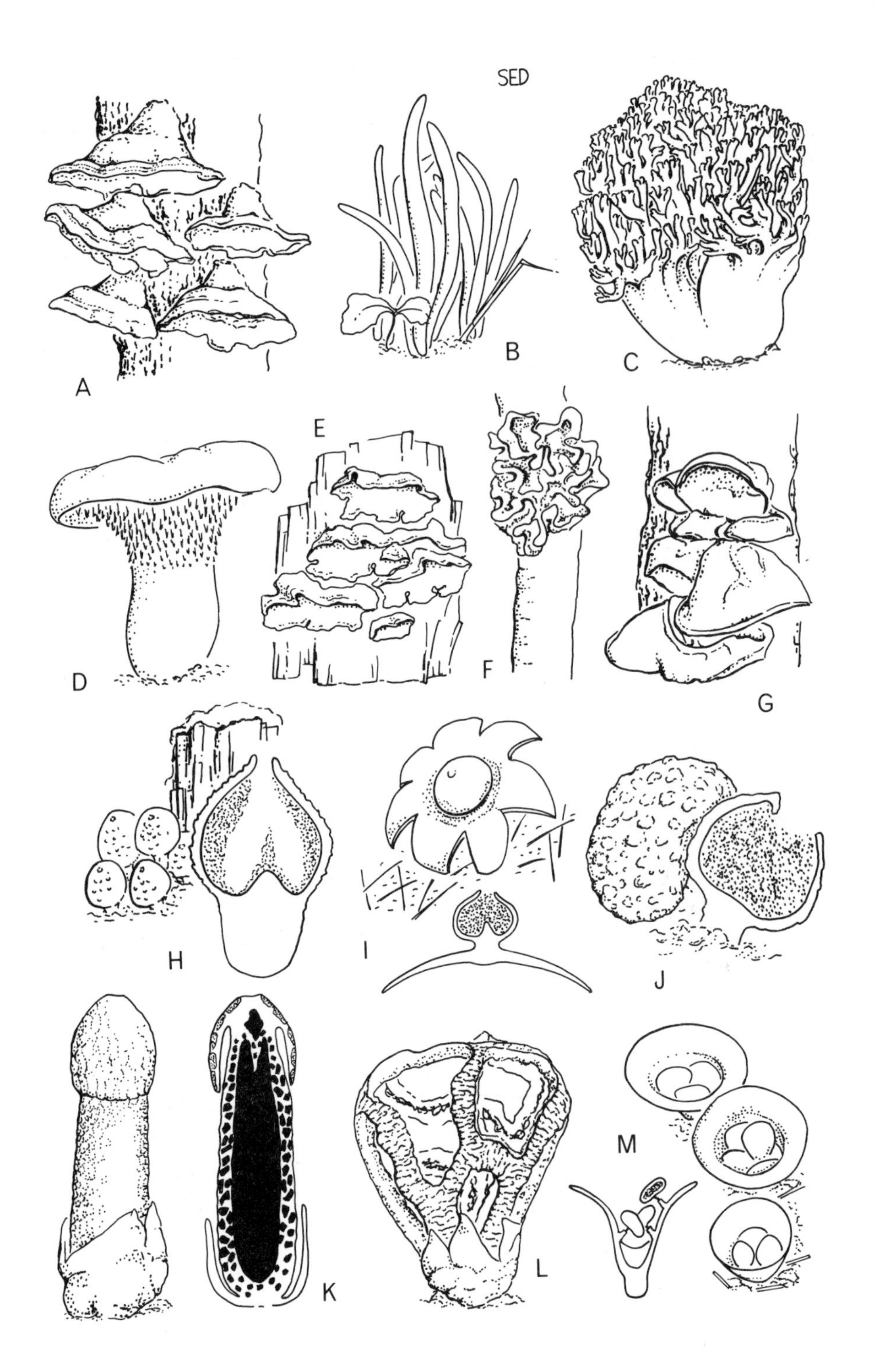

FIG. 62 — Basidiomycetes: **A** Poroid Fungi: *Ganoderma*; **B–C**, Clavarioid Fungi: **B** Fairy Clubs, *Clavaria*; **C** Coral Fungi, *Ramaria*; **D** Hydnoid Fungi: Tooth Fungus, *Hydnum*; **E** Stereoid Fungi: Skin Fungus, *Stereum*; **F–G** Jelly Fungi: **F** *Tremella*; **G** Jew's Ear Fungus, *Auricularia*; **H–M** Gasteroid Fungi: **H** Puffball, *Lycoperdon*; **I** Earthstar, *Geastrum*; **J** Earthball, *Scleroderma*; **K** Stinkhorn, *Phallus*; **L** Cage Fungus, *Clathrus*; **M** Bird's Nest Fungus, *Cyathus*.

Ascomycotina. Spores formed internally within an *ascus* (pl. *asci*) (Fig. 59B).

An extremely diverse and numerous group. The following is a very simplified guide.

1. *Cup-Fungi* (Fig. 63A–C) – Discomycetes, mainly disk- or cup-shaped fruit-bodies (apothecia). Apothecia may be greatly modified in form, e.g. in *Morchella, Helvella* and hypogeous species. Most larger types have operculate asci (thin-walled, opening by a 'lid') and belong in Pezizales. Most smaller types have inoperculate asci (thick-walled at apex, spores released through a pore), and belong to Leotiales and Ostropales.

2. *Flask-Fungi* (Fig. 63D) – Pyrenomycetes and Loculoascomycetes; asci produced within flask-shaped structures (perithecia or pseudothecia). These are small structures, rarely exceeding 1 mm. in diameter, but may be aggregated in sterile stromatic tissue to form large fruit-bodies, e.g. *Xylaria, Daldinia.* Asci may be unitunicate (Pyrenomycetes) or bitunicate (Loculoascomycetes).

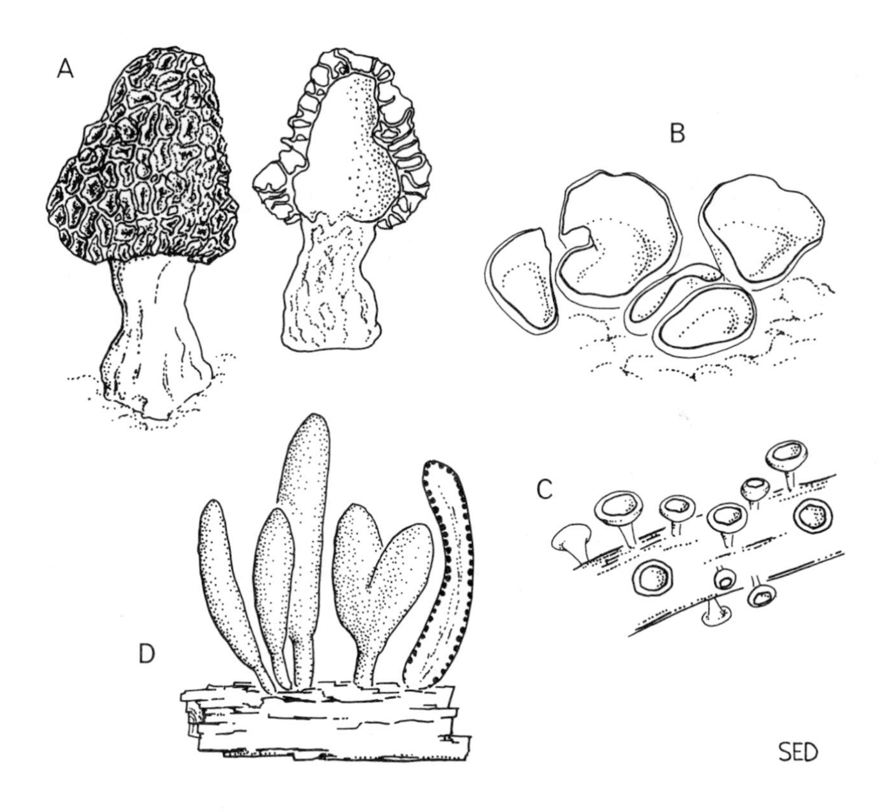

FIG. 63 — Ascomycetes: **A–C** Discomycetes: **A** Morel, *Morchella*; **B** Cup Fungi, *Peziza*; **C** Tiny Cup Fungi, *Hymenosyphus*. **D** Pyrenomycetes: Dead Man's Fingers, *Xylaria*.

COLLECTING AND PRESERVING LICHENIZED FUNGI (LICHENS)

Lichenized fungi are preserved in the same way as non-lichenized fungi. They should be air dried, then placed in envelope-type packets. With the possible exception of some foliose and fruticose species they should never be pressed. Field data should include notes indicating any reactions to chemicals exhibited by the fresh thallus.

Lichenized fungi tend to be slow-growing and sensitive to damage. Although collections should be representative of the population they are from, they should be sparing, only the minimum amount required being taken. The following points should be borne in mind:

- **Foliose** (leaf-like) species should be collected with part of the substrate and never peeled off from it, since this may cause loss of characters.

- **Fruticose** (branched) species should be collected with the basal part intact.

- **Saxicolous** species (closely appressed to rocks) should be collected using a small hammer and chisel, or knife blade if on flaky rocks, the rock type being noted. They should be separately wrapped in tissue to prevent damage.

- **Crustose** (crust-like) soil-dwelling species may be moistened with a fine mist spray if dry and brittle at the time of collection; they should be separately wrapped as above. Excess soil may be removed but sufficient must be retained to bind the thallus together; some authors recommend consolidating the soil with glue and mounting on a card.

The above is only a very brief guide but see Beatriz Coutiño 1986; Rosentreter et al. 1988; Ryan & McWhorter 1986 & Savile 1973.

COLLECTING AND PRESERVING MYXOMYCETES (SLIME MOULDS)

Myxomycetes are not fungi. They lack hyphae and have an amoeboid stage in the life cycle. However, mature fruit-bodies (sporangia) contain a powdery spore mass and are fungus-like in appearance. They are usually dealt with in mycological herbaria.

Sporangia may be sessile or stipitate, and commonly occur in swarms on various substrates, including living and dead plant parts. In some species, sporangia may be confluent, forming large structures termed aethalia. Many species contain lime in the form of granules or crystals. All myxomycete sporangia are fragile structures which must be collected carefully with part of the substrate; in the field, collections should be placed separately in small containers, or pinned onto cork or similar substrate in a box.

BRIEF GLOSSARY OF FUNGAL TERMS

agaric Any fungus having a typical mushroom-like fruit-body (fig. 60).
apothecium Cup- or disk-like fruit-body of an ascomycete.
ascus Sac-or tube-like structure, inside which spores are produced (fig. 59, B).
basidium Club-shaped structure from which spores are produced externally (fig. 59, A).
bitunicate Bounded by a double wall (refers to ascus structure).
hymenium Fertile surface; the spore-bearing layer of a fruit-body.
hypogeous Developing below ground level.
perithecium Flask-shaped ascomycete fruit-body, which contains unitunicate asci.
pileus Cap. That part of a non-resupinate fungus which bears the hymenium.
pruinose Having a powdery or floury surface covering.
pseudothecium Flask-shaped ascomycete fruit-body, which contains bitunicate asci.
resupinate Flat on the substrate, hymenium on the outer surface.
rhizoids Thread-like outgrowths from a thallus, used to anchor the plant and absorb water and nutrients.
ring (or **annulus**) The ring-like remains of the partial veil, occurring on the stalk of an agaric.
sessile Lacking a stalk (i.e. non-stipitate).
stipe A stalk; stipitate = stalked.
stroma Mass of sterile tissue on or in which spore-bearing structures are developed.
thallus Plant body.
unitunicate Bounded by a single wall (refers to ascus structure).
veil Protective layer which covers all or part of a young basidiomycete fruit-body.
volva Cup- or sac-like remains of the outer or universal veil occurring at the stipe base of an agaric (see fig. 60, A).

REFERENCES

Ainsworth et al. (1973)
Beatriz Coutiño (1986)
Cifuentes Blanco et al. (1986)
Courtecuisse (1991)
Duncan (1970)
Hawksworth (1974)
Nishida (1989)
Rosentreter et al. (1988)
Ryan & McWhorter (1986)
Savile: 58–97 (1973)

36. THE COLLECTION OF LIVING MATERIAL

Living material may be required for cultivation in botanic gardens or in gardens attached to research institutions for the following purposes:

1. Seeds for long-term storage and seed-bank distribution (see **37. Collecting for a seed bank**).

2. Seeds, plants and bulbs etc. for:

- *Research* (taxonomy, anatomy, phytochemistry and cytology).
- *Conservation* (*ex situ*) (see **41**. **Conservation and the herbarium**):

  a. To build up cultivated stock of rare plants for distribution in order to avoid collecting from endangered wild populations.
  b. Returning plants to countries of origin in order to increase stocks – with potential for replanting in the wild.
- *Teaching*: need to maintain representative collections and to distribute research material to other centres.
- *Plant introduction*: breeding and selection for ornamental use or other economic uses and subsequent release to national collections. Also for research into improvement of propagation methods e.g. tissue culture.

Nowadays the need for general collecting from a given area has largely diminished, but selective, well documented accessions can have *vital* importance. Both botanic gardens and research institutions are restricted in accommodation and resources, so guidance should first be sought in their priority requirements. It should be borne in mind that it is increasingly possible to obtain material of documented wild origin through correspondence with botanic gardens.

It is essential that collecting is carried out in accordance with national and international regulations. Local permits must be obtained if they are required. But even in the absence of local laws protecting the vegetation, collectors must exercise restraint and take care not to endanger the survival of any rare plants in the wild. The careful collection of endangered plant material can enhance a species' long-term chances of survival; on the other hand, the reckless harvesting of a localized endemic could result in its extinction.

When collecting plants, all stages of the process must be carried out in full cooperation with the host authorities. All legislation regarding plant collecting and exportation should be carefully scrutinized, remembering that germplasm of potential crop genera could be a source of foreign exchange and therefore export may be prohibited or tightly controlled. When negotiating permission to collect, the proposed subjects and their usage should be stated and, if required, the promise of duplicates made and honoured. Record the status of the land, e.g. nature reserve, and source of permission, for reference by future collectors in the area.

If collecting in another country, a local botanic garden should be contacted and if possible visited in order to establish goodwill and to seek co-operation. Any desired material not collected could possibly be provided by the garden at a later date.

Check prior to collecting whether the plants are prohibited or restricted items for import into the destination country. (see **41. Conservation and the herbarium**, *CITES*).

LOGISTIC CONSIDERATIONS

The collection of viable material may pose problems additional to those encountered for preserved specimens:

- Communication and travel systems must be carefully checked; the type and speed of transport will affect any collecting strategies. When collecting high montane species for example, it is imperative that they are moved through any tropical lowland areas as rapidly as possible.

- Living non-seed material is irregularly shaped and will require careful handling; it can be bulky, so adding to transport and packing difficulties.

- When routes have been organised, arrangements should be made for temporary nurseries and holding grounds or for shipping points during expeditions.

FIELD-NOTES

The data to be recorded are as for herbarium material (see **30. Collecting and preserving specimens**) except that additional information which might be helpful for cultivation must be included, especially:

Altitude. This is of critical value in mountainous areas where local climatic gradients are steep and narrow.

Habitat. Detailed notes are of great importance in identifying the precise cultivation requirements of collections. The following in particular should be recorded in greater detail than is necessary for preserved specimens:

- Soil type: acidity/alkalinity (give pH value if possible), consistency and depth, and underlying rock type.

- The local micro-habitat should be noted in addition to the general vegetation type (e.g. growing only in open glades within short grass sward).

- Aspect (e.g. growing in full sunlight or shade or in exposed or sheltered situations).

- Diurnal variations: are there differences of temperature or humidity between day and night? This can be extreme in alpine regions and minimal in tropical lowlands.

- Seasonal climatic variations: are there annual periods of floods, drought or frosty conditions? These are often difficult to assess without local experience.

Population notes

Special collections for breeding programmes may require population notes; usually these projects will only be undertaken with expert guidance and cards provided listing characters to be noted. For example, one may be required to look for differences between individual plants for resistance to disease, general vigour or abundance of flowers or fruit. When collecting for a conservation programme it is recommended that the guidelines laid down by the Center for Plant Conservation are followed (see Center for Plant Conservation 1986).

COLLECTING TECHNIQUES

Wherever possible collect a fertile voucher herbarium specimen for each living specimen.

Seed is often the most convenient and successful way of collecting and transporting plants, as seed is adapted for dispersal and is thus resistant to many of the stresses of collecting and transport (see also **37. Collecting seed for the seed bank**). The quantity of material collected must not place strain upon local populations; ideally either seed or perennating organs (e.g. bulbs, rhizomes, corms etc.) rather than entire plants should be collected. By selecting perennating organs or propagules, transport is easier and the probability of successful transport and establishment increased.

Plant collecting can be split into a number of different stages: locating the plant, physically collecting it, maintaining it during the collecting trip and finally dispatching it. Each stage requires care and a lot of effort; collecting live plants can be more time-consuming than the collection of conventional herbarium specimens.

Having located the plant, examine it closely; avoid collecting obviously sick or diseased specimens. Choose the most appropriate collecting techniques: bulbous plants can be dug up using a hand-fork or spade, while aerial parts are cut with secateurs or a sharp knife. Select the best part of the plant to use, e.g. laterally rooted stems, aerial roots on stems, rooted suckers and runners etc. Carefully lift from the soil and remove unwanted damaged leaves etc.

Most plants can be stored prior to shipping wrapped in damp newspaper with the leaves open to the air. If there is a risk of desiccation, the roots can be wrapped in damp moss and then wrapped within a plastic bag. If entirely closed in plastic bags without free ventilation, the high humidities and fluctuating temperatures will rapidly rot living tissues. The packages can be packed into open boxes and stored in vehicles. Even in temperate regions, temperature fluctuations within vehicles can be very damaging to collections; ventilate when there is no risk of dust and frequently check and moisten plants. An insulated cool box is recommended when transporting very delicate material such as aquatics or montane material. If the plants are packed early in the morning the lethal extremes of midday are avoided; in the evening the plants should be unpacked and checked for desiccation or rotting. Material must be inspected frequently and leaves or tissue showing signs of rot must be removed. Rotting may well signify that the plants are being kept too wet; they should be kept merely moist.

Notes on special groups

Bulbs and Rhizomes. Easy to collect although sometimes requiring some strenuous digging. Store in stout manila envelopes and allow the foliage to die down, then pack dry in sawdust (entry prohibited to some countries) or polystyrene chips. Do not pack in plastic bags as the bulbs will 'sweat' and rot.

Seedlings and divisions. A number of taxa are easily collected as either seedlings or as rooted divisions from the parent plant; packed carefully into moist moss or leaf-litter within an open plastic bag they will generally transport well. Rooted runners, stolons and stems, e.g. Araceae, can be transported in the same way.

Cuttings. When plants can be collected and rapidly transported it is convenient to collect cutting material. Taking hardwood cuttings of temperate and some tropical shrubs, e.g. *Bougainvillea* and some members of the Burseraceae, is one of the cheapest and easiest methods of propagation. When collecting during the dormant season, choose woody shoots of moderate vigour and size from the previous season's growth. If unsure what type of material to collect, choose shoots with a section of older wood attached in the form of heel or mallet. Sealing the ends of cuttings with paraffin wax or candle wax will prolong transport life.

Softwood cuttings are not so resilient; they can be dispatched by air if wrapped in moist newspaper within a padded envelope provided they are in transit for no more than two days, otherwise the material will suffer.

Aquatics can be split into two broad groups, those with a basal perennating organ that can be transported with relative ease, and those with delicate foliage and roots that require more careful transport.

Large tuberous taxa such as *Nelumbo* and *Nymphaea* have basal rhizomes or tubers. These should have the foliage removed, with the tuber subsequently stored for transport in slightly damp peat or moss. Other taxa such as *Echinodorus* or *Cryptocoryne* have a smaller basal organ and the leaves should be retained; the roots should be wrapped in damp moss within a plastic bag, and the whole plant rolled in newspaper. Where possible collect individuals of the required species growing in a marginal position with the leaves out of water; they will be far easier to transport.

More delicate aquatics such as *Cabomba* or *Myriophyllum* should be wrapped in sheets of damp newspaper which is regularly moistened and kept cool. Transit should be as rapid as possible.

Succulents. Many succulent taxa (e.g. Cactaceae, all *Aloe* spp. and succulent *Euphorbia* spp.) are governed by CITES regulations. Many cacti are now endangered through the removal of plants from the wild for the horticultural industry, so priority should always be given to the collection of seed.

Succulents are adapted to drought and therefore can easily survive collecting and transport. Cactus seedlings and also cuttings can be wrapped around with dry newspaper or packed dry in sawdust. The outer leaves of *Aloe* plants or cuttings can be cut back severely without harm. They can be stored in cloth bags or wrapped in dry newspaper for weeks, as can the bulbils of members of the Agavaceae. Stem cuttings taken from *Yucca* and leaf cuttings from *Sansevieria* are resistant to desiccation and can be stored dry in cloth bags or envelopes. *Euphorbia* cuttings or small plants with roots intact can be stored in dry newspaper for several weeks. Many arid-land bromeliads can be collected in the same way as succulents.

Ferns. Pack the roots in a convenient moist medium and place into an open plastic bag and pack around with moist newspaper. Select small juvenile plants since these adapt to cultivation more easily, or use rooted sections of rhizome. Spores can be collected for germination; see Roos & Verduyn (1989).

Orchids. As for ferns but orchids are quite resistant to periods of drought and will transport well; a moist medium is therefore not always required. Epiphytic taxa are more tolerant to dry conditions than terrestrial taxa. Terrestrial taxa are best collected when flowering and then stored in a damp medium.

DISPATCH BY AIR

The dispatch of plants internationally by air-freight is recommended despite its great expense; it is comparatively rapid and, what is important, parcels can be traced. When packing select a stout cardboard box and line with sheets of polystyrene foam. Plants for dispatch must be checked by the appropriate plant health authorities who may have to issue a phytosanitary certificate. Excess leaves are taken off and all soil removed from the roots. The plants are packed tightly between sheets of moist newspaper (or dry, for succulents) and if necessary the roots wrapped in a plastic bag containing damp sawdust, peat or polystyrene chips. The plants should be tightly packed to stop them moving. Heavy items should be packed low and light items high in the box, which should be labelled clearly to indicate the right way up.

Within the box include the sender's name and address, collector's name, numbers and other relevant data in a sealed and waterproof plastic bag. In case of loss it may prove useful to retain a back-up copy of the notes. All plants should have adequate labels bearing name and collecting number firmly attached to the plant material. The parcels must all be clearly marked on the outside, as follows: 'LIVING BOTANICAL SPECIMENS', 'STORE IN PRESSURIZED COMPARTMENT' and 'NO COMMERCIAL VALUE'.

The sender must be aware of any regulations concerning the import of plants into the country of destination and must make any necessary arrangements. Some botanic gardens are officially recognized as quarantine stations, in which case any consignments addressed to the garden are freely allowed into the country.

REFERENCES

Archer: 40–43 (1945)
Balick (1989)
Center for Plant Conservation (1986)
IBPGR (1983)
Roos & Verduyn (1989)

37. COLLECTING FOR A SEED BANK

Aims of seed collecting policy

The aims of a wild species seed bank might be defined as the establishment for long-term storage of well documented and verified collections of viable (living) seed of wild species. The seed is carefully dried to low moisture content and then stored at low temperatures. Each sample represents, as far as possible, the genetic variation within the population from which it was taken. The collections then act as a basis for conservation or as a source of material to *bona fide* institutions for all aspects of biological study.

Permission for seed collecting

Many countries now have specific legislation covering the collection and export of seed and other plant material. It is the collector's responsibility to ensure that the proper export permits have been obtained before commencing collection (see Balick: 479 (1989)).

International codes of conduct for seed collecting are being debated at the FAO Commission on Plant Genetic Resources and the Biological Diversity Convention of UNCED (United Nations Conference on the Environment and Development). Their decisions will set out the procedures which should be followed.

Once conserved in the bank, samples of larger collections may be offered to research workers worldwide. However, representative samples of everything collected should also be deposited in the appropriate national institution of the host country.

Collection size

The size of the collection determines what the bank is able to do with it. For the collection to be of greatest benefit there should be enough seed both for long-term conservation and distribution to research institutes. Hence a collection must have at least 1000 seeds to be useful for conservation alone, 5000 or more is tolerable and allows both conservation and some distribution, while 20,000+ is ideal, ensuring that the seed stocks will be sufficient to last for the centuries that the seeds should remain viable.

Smaller collections should only be contemplated where the material is of special scientific interest. Collections of less than 500 seeds are a problem to bank. As viability declines, the necessary growing on of the seed to form a basis for a future seed crop becomes unpredictable and success cannot be guaranteed. Only when the species is endangered or endemic and only a few plants can be found are collections of less than 500 seeds acceptable for inclusion in a bank.

Sampling in the field

Seeds should be gathered evenly and at random from as many individuals of the parent population as possible.

In the case of annuals, biennials and short-lived perennials, no more than 20% of the available seed should be taken from each plant. With long-lived perennials, e.g. trees and shrubs, this restriction can be applied less strictly.

It should be noted that within a morphologically uniform stand of plants substantial genetic variation in the physiological responses of the individuals can be hidden. An example of this hidden variation is demonstrated by plants growing very close to stream edges compared with those growing further away. When both are grown under identical conditions they can differ substantially in their physiological behaviour. When sampling, the plants growing by the stream edge should be considered as a distinct population from those growing a distance away. Likewise plants growing on well-trodden footpaths through a field are known to differ substantially in their flowering behaviour from the individuals of the same species growing either side of the path. Therefore, when collecting the plants on a path they should be considered as one population and those growing either side of it and in the rest of the field to be another.

Before harvesting, a sample should be examined for the presence of fully formed seeds. The entire seed-heads can be collected. There is no need for seed cleaning.

Collecting data and pressed voucher specimens

The data which should be recorded are similar to those required for herbarium specimens (see **30. Collecting and preserving specimens**). In addition, the number of plants found and sampled should be noted so that others using the material will have an indication as to how well the population was sampled. If the seed was collected from the ground, this should also be noted.

A pressed herbarium voucher specimen should be prepared so that the identity of the seed material collected can be verified. It should be numbered to correspond with the seed sample. Where possible the specimen should include flowers, fruits and representative vegetative structures (including the roots on annuals etc.). The location where the voucher is to be preserved should be recorded. If no voucher is prepared, this fact should be noted.

Suitable species of families for collection

With present technology not all seeds are suitable for storage in seed banks. Some seeds, especially large fleshy ones and those of some aquatic species, lose viability on drying; these seeds, or the species to which they belong, are known as *recalcitrant* and should not be collected.

Families of primary interest for many seed banks will be:

Amaranthaceae	Cruciferae	Gramineae
Chenopodiaceae	Cucurbitaceae	Leguminosae
Compositae	Euphorbiaceae	Solanaceae

Ensuring maximum longevity for the seed lot

Prior to collection a sample of the seed in each population should be removed from the seed-heads, capsules or fruits and checked for fullness using forceps. Obvious signs of insect infestation should also be looked for. If a small proportion of the collected seed shows signs of infestation the insects will die when the seed is dried and frozen. However, badly infested populations should be avoided.

The seed should be placed in well secured cotton bags. Polythene bags should **not** be used unless moist berries are collected, and in this case the bags should be occasionally opened for aeration. All bags should be clearly labelled both inside and outside with the collection number written in pencil on a number tag. They should be left to dry at ambient temperature and humidity, away from direct sunlight and then packaged carefully for shipment.

Seed treatments such as fumigation, application of dressing or even UV irradiation should be avoided as they may seriously affect storage life. Only where the authorities insist should treatment be allowed.

Because the length of time between collection and storage in the bank is critical and delays can seriously affect the seed's viability, every effort should be made to transport seed to the seed bank by the quickest route. While in transit it is vital to keep the seed as cool as possible (20°C or lower, but do not freeze) and also well ventilated. When using air-freight the Air Way Bill (AWB) number should be sent to the intended recipient so that they can monitor its progress.

REFERENCES

Balick (1989)
IBPGR (1983)

38. PHOTOGRAPHY AND FIELDWORK

Photography is a useful and sometimes vital adjunct to other methods of collecting data in the field. Good photographs (whether prints or colour slides) can provide a good deal of useful information, besides being a permanent record of a particular plant or habitat.

BASIC EQUIPMENT

Camera. A 35 mm. single-lens reflex (SLR) camera is suitable for most purposes and particularly for close-up work, allowing one to see precisely what one is going to photograph. A built-in light meter is a part of most modern cameras and is particularly useful for those with little photographic experience, making it relatively easy to estimate the exposure for any photograph. Cameras with many options are to be avoided as they can be too complicated to use under field conditions, are expensive to buy and to repair when they go wrong, and are really only suitable for experienced photographers.

Lenses. Only 3 lenses are generally required – a standard lens (normally provided with the camera), a macro-lens (e.g. 50 or 80 mm. is most suitable), and a zoom lens (e.g. 70–200 mm. or similar). Too many lenses will be difficult to carry in the field. A good macro-lens is absolutely essential for close-ups of flowers, fruits etc. Most modern macro lenses of approximately 50 mm. can also be used instead of standard lenses, further reducing the weight of equipment in the field. An alternative is a zoom lens with the macro facility built in, but this leads to some loss of quality in the pictures.

Filters. The use of filters is a complicated subject and they can only be recommended to those with some experience of photography. However, standard UV filters or skylight filters are useful; they cut down haze and protect the lens.

Light meters. Most modern cameras will have a built-in light meter. However, some users prefer a separate, hand-held light meter. Whichever is used, it is essential that it is proved to be reliable and working properly before use. It is an advantage to have both types so that occasional checks on readings can be made to ensure that the meters are functioning correctly. The basic function of a light meter is to provide a scale of exposure readings for a given film speed; the shutter speed at which the picture is taken is set against the lens aperture, the ideal in general being to try and get as small an aperture and as fast a speed as possible.

Tripod. An optional piece of equipment, but essential if a lot of close-up work is contemplated. A good tripod holds the camera absolutely steady, allows a picture to be composed carefully and also allows time exposure (i.e. slow shutter speeds that cannot be hand-held) to be used. Light weight tripods are available for field work.

Flash. The use of electronic flash units to aid photography in difficult conditions (in deep shade, in a forest and so on) is another useful adjunct, especially if photographs are wanted only for the record. It is difficult to get natural-looking photographs with any flash unit unless it is used with great care. However, the fine range of modern high-speed films, together with the use of a tripod, in many cases render the use of flash unnecessary. Flash units designed to work with a specific make of camera (dedicated flash) are highly recommended.

Camera bag. A good lightweight camera bag, or a rucksack with fitted compartments for lenses etc., is essential.

Film. The use of black and white film, colour print film or transparencies is a matter of individual choice. Transparencies generally have the advantage that they can be used for lecture purposes, and prints can, if required, be taken from them. Make of film again depends really on the preferences of the individual. Most (e.g. Kodak, Fuji, Agfa) have a range of colour transparency films with speeds from ISO (or ASA) 64, 100, 200 to 400 or more. On the whole, an ISO of 64 or 100 is adequate for most purposes, but under forest conditions 200 or 400 is preferable. Kodochrome 25, 64 and 200 ISO are highly recommended for field work particularly in areas of high or low temperature and when speedy processing is not possible. Generally, the faster the film the more grainy the picture and the less good the colour balance. Prints are useful to add to herbarium sheets or illustration collections.

Field notebook. Absolutely essential for any organised scientific work. Each photograph taken should be noted down giving it a number (each film numbered 1–36 for colour transparencies), date and as precise a geographical locality as possible. The subject should be noted, e.g. view, whole plants, close-up and so on. It is also important to record if a herbarium specimen has been made or seed sample taken etc, and to make sure that both are cross-referenced in the appropriate notebooks. It is often advisable to keep a copy of notebooks in case one of them gets lost.

Other equipment. Waterproof containers, polythene bags and silica gel are very useful to prevent cameras and films getting damp. If working in humid conditions it is wise to wrap all equipment into sealed polythene bags, but not if they are already wet. In such a situation equipment must be first dried before putting it in a bag. An X-ray proof bag for photographic film is also useful – some airports do not have film-safe X-ray machines. Film must always be carried in hand-luggage when flying.

TAKING PHOTOGRAPHS.

Knowledge of how your camera works is essential before taking any photos. Before proceeding, read carefully the manual which comes with the camera.

In the field a decision has to be made exactly what to photograph, i.e. habitat, individual plant or close-up, or all three. Remember that distant habitat shots or distant shots of bushes or trees are rarely of much use. Good, clear, reasonably close landscape-habitat shots are very useful (though not many of each will be wanted), especially if taken from a raised view-point, e.g. standing on a rock or vehicle bonnet.

Close-ups of flowers, fruits or leaves are very useful, especially to record details that will be lost when specimens are dried (colour, shape etc). Close-ups of flowers and fruits must be large enough so that the desired details can be clearly seen.

Use a tripod whenever possible. Compose the picture and focus with care. SLR cameras have a built-in screen usually with split-image range finder to achieve critical focussing.

The most critical part of taking a photograph is choosing the correct shutter speed. First it is essential to make sure that the light meter is set for the ISO (or ASA) film speed. Hand-held shots should be no longer than 1/60 second, preferably not lower than 1/125. High speeds such as 1/250, or even 1/500 will be required if the conditions are windy or the subject is moving (such as an animal, pollinator etc).

If a tripod is used and the conditions are still, then much slower speeds can be employed to advantage; allowing smaller lens apertures. Remember that the smaller the lens aperture (the greater the f-stop number) the more of the depth of subject will be in focus. This is the whole essence of taking a good, clear photograph.

On the whole it is wise to take at least 2 photographs of each subject, but remember that taking excessive numbers of photographs can be expensive.

When changing film under adverse conditions (wet or dusty), care must be taken not to get moisture or dust into the camera. If possible change films with the camera inside a plastic bag or in a closed tent etc.

AFTER CARE.

When films have been taken, seal them in a waterproof cartridge (provided with the film) and place them inside a sealed bag for extra protection. Keep them away from heat and moisture. If a refrigerator is at hand, films (exposed or unexposed) can be stored for months without any deterioration. However, before inserting in the camera a film should be allowed to warm up to ambient temperature to prevent condensation forming.

It is most important that films should always be developed at the first opportunity.

Once the pictures (prints or slides) have been returned from processing, each film should be marked with its field number (film number 1 etc.). Each picture then needs to be identified and labelled on the slide, or back of print, with essential information from the field notebooks. Most critical is the collector's number, which must be noted on the photograph if there is a corresponding herbarium or seed specimen. Other data which need to go on the photograph are the plant's name, country and location, date and photographer's name.

Prints can be incorporated onto the appropriate herbarium sheet, or filed in albums or the illustrations collection. Slides can be stored in sleeves or boxes. There is no perfect way of filing slides, but a numerical sequence seems to suit most people. A card-index (alphabetical by genus) enables any slide to be quickly located (see **14. Collections of illustrations and photographs**).

All photographs need to be handled with care and stored away from heat, damp, dust and daylight. Remember, if properly looked after they are a lasting record and can be used for many years to come for lecture purposes and publications. In some instances, they may represent views of habitats that have been destroyed or species that have become extinct.

REFERENCES

Blacklock & Blacklock (1987)

39. ECONOMIC BOTANY AND THE HERBARIUM

Mankind is completely dependant on plants for its existence. All our food comes either directly from plants or is derived indirectly from them through the animals we eat. Similarly, our clothing and many other products we use are either of plant origin or are made from animals which depend on plants for their existence. A major plant product we use in many ways is wood. Our fossil fuel supplies (coal, oil and natural gas) are derived from ancient plant remains. Thus most of our energy needs, both food and fuel, rely on the ability of plants to capture energy from the sun's rays by the process of photosynthesis, enabling them to grow and reproduce. Knowledge of the uses of plants is therefore essential to human progress.

Economic botany is concerned with:

- Recording the uses and values of plants to mankind.
- Investigating variation in useful plants, e.g. their yield and quality, with a view to realizing their potential in breeding programmes.
- Promoting the use of indigenous plants and, if appropriate, the introduction of useful plants to different regions and societies.
- Supplying data which could lead to the development of new plant products, e.g. pharmaceuticals, pesticides, etc.
- Ensuring the sustainable use of wild plants, i.e. leaving sufficient plants to continue growing, regenerating or reproducing for future use.
- Consideration of the use of alternative species where over-exploitation has occurred.

USES OF PLANTS

A standard database format for recording data on the use of plants is currently being devised at Kew in discussion with botanists from other countries. This standard may be ratified by IUBS - TDWG (International Union for Biological Sciences - International Working Group on Taxonomic Databases for Plant Sciences) in November 1992. The categories within this scheme will also be of value to those wishing to adopt standard terms for uses on specimen data labels. The major categories proposed are:

- *Food,* including beverages and food additives.
- *Animal food,* forage and fodder.
- *Bee plants,* pollen or nectar sources for honey production.

- *Host plants,* including the food-plants of useful organisms such as silkworms, lac insects and edible grubs, and hosts of pests and diseases of major crops.

- *Materials,* including woods, fibres, cork, cane, tannins, latex, resins, gums, waxes, oils, lipids etc.

- *Fuels,* including wood, charcoal, petroleum substitutes, fuel alcohols etc.

- *Medicines,* both human and veterinary.

- *Social products,* i.e. those used for social purposes not definable as food or medicines, including masticatories, smoking materials, narcotics, hallucinogens and psychoactive drugs, contraceptives and abortifacients.

- *Poisons,* e.g. molluscicides, insecticides, fish poisons etc.; also accidental poisons, e.g. to livestock.

- *Land uses,* e.g. intercrops and nursecrops (i.e. plants grown in between the main crop to provide a secondary crop and/or shade while the main crop matures), ornamentals, barrier hedges, shade plants, windbreaks, soil improvers, plants for revegetation and erosion control, waste water purifiers (i.e. plants such as *Phragmites* which act as living filters and also absorb impurities) and indicators of the presence of metals, pollution, underground water etc.

- *Gene sources,* the wild relatives of major crops and **landraces** (i.e. the ancient or primitive cultivars of crop plants) which may possess useful qualities, such as disease resistance, cold hardiness etc., of possible value in crop breeding programmes.

- *Weeds,* which cause losses to crops and livestock production.

COLLECTIONS

General herbarium material and ancillary collections can be a major source of information on plant use. Much economic information can be taken from data labels, and important reference works have been compiled largely from such sources (e.g. Burkill 1985). The specimens themselves can provide valuable morphological data (e.g. fruit and seed size) and phenological data (i.e. flowering, fruiting times) to the economic botanist.

Special economic botany collections supplement standard herbarium collections. These may contain material better described as museum items, e.g. collections of:

- plant products such as pigments, gums, lacquers and oils.

- timbers and wooden artefacts.

- plant fibres and artefacts made from them, e.g. paper, textiles, string, baskets.

- archaeological plant materials.

In addition, special collections of specimens, both herbarium and ancillary, may be built up according to uses, e.g.:

- food plants.

- medicinal plants.

- poisonous plants.

Economic botany collections may have an international, national, regional or local representation. They can be used as comparative identification aids and, if regional or local, form a base from which check-lists of useful plants can be compiled. They can also serve as a source of material for analyses (e.g. chemical, anatomical etc.).

VOUCHER SPECIMENS

Voucher specimens deposited in national or local herbaria are vital to studies of economic botany because they enable information to be positively associated with specimens whose identity can be verified. Hence the information gathered can be reliably attributed to that species and, if future doubt arises, can be checked. Surprisingly, many studies in economic botany, including those which involve expensive and time-consuming laboratory research, are seriously weakened by neglect of this important requirement (Lewis et al. 1988).

Collecting voucher specimens

Collecting voucher specimens – (see also chapters **30**, **31**, **36**, & **37**).

Voucher specimens usually take the form of dry herbarium specimens and/or ancillary collections (e.g. spirit, carpological and wood samples). However, if inadequate for identification, it may be possible, where appropriate, for propagating material to be collected and, after cultivation, a voucher prepared. Living vouchers may also be collected where appropriate and maintained in botanic gardens. Whenever possible, collections should be accompanied by photographic records. All notebooks and transcripts should be kept together under safe archival conditions with the subsequent results of collaborative studies, e.g. chemical analyses etc.

Size of collection

At least one good voucher specimen, and sufficient material for analytical study will be needed.

Collection of duplicate vouchers as well as material for analysis is desirable but not always feasible in ethnobotanical studies (see Lewis et al. 1988 and Pake 1987). Very often the cooperation of a community depends on not appearing excessively demanding; for example, if collections are from home gardens excessive collecting could denude resources.

Multiple collections

Lewis et al. (1988) suggest collection of the same species at another locality or time of year for different stages of development to be represented in order to aid identification. Multiple collections of the same species also help to establish the frequency and consistency of use from community to community. In the case of medicinal plant use, consistent use can also indicate effectiveness. Multiple collections also serve as a check on folk identification.

Collecting from markets

Markets are excellent sources of information on economic botany. Methods applicable to market collection are described by Bye & Linares (1983). The recommended method for obtaining vouchers for the study of fresh plants sold as medicines and food is to buy the standard units and prepare herbarium (or spirit) specimens from them. In the case of seeds, stems and roots, material may be bought for propagation.

In addition to vouchers it is important to purchase samples of plant material for analysis. Photographic records are most valuable. Check if the vendor is a resale vendor or collector vendor and try to ascertain the origin of the material; follow with field trips if appropriate.

Recording data for voucher specimens

The quality and integrity of the data recorded are of great significance in vouchers for economic studies.

Apart from the usual data (see **30. Collecting and preserving specimens**, *Recording data*) it is necessary to collect additional kinds of data to accompany economic collections; these can be divided into *general* and *uses.*

General data, record:

- *Source of information* on use (this may not be the same as the user). It is important to indicate if the information is from personal observation or has been otherwise gathered. In the latter case it will be necessary to record details on the informant (name if appropriate, gender, approximate age, occupation, ethnic identity, locality and language spoken).

- *Types of use* – the purpose for which the plant is used, and the importance of the use or product.
- *Plant part* used, distinguish whether mature or immature.
- *Production* – details of harvesting, processing and storage; take note of any special tools used.
- *Users* – specify: ethnic group (tribe/caste/sect etc.), from which locality or area, and whether the use is traditional, modern or potential.
- *Vernacular name(s)* – the name of the plant must be distinguished from the names of plant parts, the names of any processing stages, and the final product. The use of some names may be restricted to the time of year or occasion. The language and dialect should be stated, and care taken with transliteration into the Roman alphabet.
- *Season of availability and of use* and, if relevant, time of day and method of storage.
- *Ecological data*, it is important to record all data which might be relevant to the future cultivation of the species – see **36. The collection of living material**.
- *Conservation data* – if from wild source record frequency, details of regeneration, sustainability or over-exploitation. Note any local restrictions.
- *Cultivation methods*, if relevant, record local methods of cultivation and harvesting, also note yield and pests and diseases.
- *Economics* – is there trade in the plants (e.g. erosion control plants, bee plants, house plants) or plant products, and on what scale?
- *Popularity* – is the plant highly valued for its use or are other species preferred and if so, which? Try to record any factors influencing the preference, e.g. rarity, inaccessibility, depletion of the resource or if it is of low quality or difficult to harvest or process.
- *Problems* – is the plant an invasive crop weed, a host for pests or diseases, or a livestock poison?
- *Potential* – could the use, cultivation, production of product etc. be increased and/or introduced to other areas?

Data on uses, record:

Food

- Food class, e.g. staple, snack, ceremonial food or famine food etc.

- Basic food type, e.g. vegetable, cereal, food additive, e.g. flavouring, sweetener, fermentation agent, etc.

- Preparation/s: e.g. used in beer, bread, soups etc.

- Processing details – how is it cooked, preserved or detoxified? If collected from a market the information should be corroborated with both vendors and purchasers.

- Qualities and values – palatability, flavour, nutritional/antinutritional value (distinguish between actual or supposed).

Animal food

- Which species of animal feed on the plant? Avoid general statements such as 'eaten by livestock/animals'.

- When eaten: all year, seasonal or in time of adverse condition only? If season of availability and season of use differ, how is the fodder stored?

Materials

- Specialized uses and products – e.g. wood used for construction, furniture, tool-handles etc.; volatile oils in perfumes etc.

- Properties – e.g. for wood: weight, hardness, colour, graining, size, durability, workability etc.; for fibres: fibre-length.

Fuels

- Fuel type – wood, charcoal, oil etc., or more than one.

- Specialized use – cooking, heating or smelting etc.

- Harvesting – is it taken green, i.e. either coppiced (cut at ground level) or pollarded (cut well above ground level), or is it gathered dead? Note also quantities and frequency taken and regeneration abilities of the plant.

- Qualities – how much heat and smoke are generated and how rapidly does it burn?

Medicines

- Symptoms or diseases treated.

- Type of medicine, e.g. antidote, prophylactic, palliative etc.

- Method of processing and recipe.

- Method of administration – internally (oral ingestion, inhalant, suppository, injection) or externally (ointment, wash, drops, poultices etc.), or as charms.

- For veterinary uses state which animal(s) are treated.

Social products

- Use and reason for use – e.g. masticatories, smoking materials or hallucinogens may be used for recreational or ritualistic purposes.

- Method of processing, recipe and any restrictions on time of use.

- Effectiveness; how long does the effect last?

- Is provenance important?

Poisons

- Organism affected, including humans, animals, plants, microorganisms.

- Uses of poisons – record specific uses, e.g. molluscicides for disease vector control; fish and arrow poisons for hunting; fungicides for protection of stored products; herbicides for weed control etc.

- Symptoms caused by the poison.

- For accidental poisons – record animal species affected (including humans) and whether the poisonous properties are dependent on other factors such as exposure (of the animal) to sunlight.

Land use

- Specific environmental data – e.g. utilized for tolerance of particular conditions; will it grow, for example, on sand dunes, deserts, watercourses, mine wastes or urban streets?

- Invasiveness – could it become a troublesome weed? Does it produce runners, deeply penetrating roots or large quantities of seed, or is it eaten, and thus controlled, by animals?

Gene Sources

- Specify related crop species.

- Note useful traits – e.g. a short stout stem for cereals, large fruit, disease resistance etc.

REFERENCES

Archer (1945)
Burkill (1985)
Bye & Linares (1983)
Cook & Hastings (1991)
Fosberg (1960)
Lewis et al. (1988)
Pake (1987)

40. ECOLOGY AND THE HERBARIUM

Ecological information is of direct importance to taxonomic studies. Plant collectors should therefore know enough about ecology to be able to make accurate descriptions of habitats for their specimens. Ecologists rely on herbaria for various data and especially for the identification of the plants making up the vegetation they are studying. On the other hand, taxonomists need ecological information for their work.

ECOLOGICAL INFORMATION AND HERBARIUM LABELS

Herbarium labels must bear at least some information about the habitat of a plant. This information helps anyone wishing to find the plant to be aware of the likely habitat, and it may also be useful in conservation and management. The habitat may be characteristic of a species, helping to distinguish it from closely allied species.

There are two aspects to the description of habitats: the general vegetation type, and other environmental factors.

The general vegetation type

Two types of classification are available:

- *Physiognomic*: using the general or characteristic form of the vegetation, e.g. grassland, woodland etc.

- *Species-based:* relying on the occurrence of certain common species to define the vegetation type.

Physiognomic classifications are usually most useful internationally; species-based classifications may be useful within a country.

If a continent-wide classification scheme is available, this should be used. Examples are White 1983 (for Africa), Eiten 1983 (for South America), Steenis 1957a (for Malesia) – these are all physiognomic classifications. A world-wide physiognomic classification exists (Unesco 1973), but has not achieved the wide acceptance that it probably deserves.

If there is no continent-wide classification scheme, the use of a local one, if available, should be considered. Such a scheme should be used if it is locally accepted and has some international acceptance. If it is species-based, try to expand it by adding physiognomic terms which will make the categories more widely understood internationally (e.g. '*Isoberlinia doka* woodland' or '*Vossia cuspidata* seasonally flooded grassland').

Avoid sole use of vernacular terms of only local significance; terms such as *miombo, fadama, cerrado, mallee,* etc., may be clear locally but mean nothing elsewhere.

Main physiognomic categories (After White 1983):

- *Forest*: woody, multi-layered, no continuous grass layer.
- *Woodland*: woody, generally one tree layer, grass layer present. Tree canopy cover greater than 40%.
- *Bushland and thicket*: woody; bushes and small trees, 3–7 m. tall; grass may be present.
- *Shrubland*: woody; shrubs up to 2 m. tall. Not much grass.
- *Grassland*: grasses or other herbs. Scattered trees or bushes may be present, cover less than 10%.
- *Wooded grassland*: grasses or other herbs; tree canopy cover 10–40%.

Minor categories - partly physiognomic and partly habitat-based (after White 1983):

- Mangrove.
- Strand and seashore vegetation.
- Desert.
- Freshwater swamp and aquatic vegetation.
- Saline and brackish swamp.
- Bamboo.
- Disturbed habitats (crop weed communities; roadsides; waste ground, abandoned farms, etc.).
- Montane vegetation.

Other environmental factors

Soils

The classification and description of soils is complex, and there is no internationally agreed standard. The two most useful features to record are colour and texture.

Colour is best described when the soil is wet. For detailed work, Munsell Soil Colour Charts can be used, but a subjective colour description is much better than nothing at all.

The texture of a soil is determined by the relative proportions of particle size classes in its mineral matter; in ascending order of size, the standard classes are clay, silt, fine sand, coarse sand, and gravel. A soil with a mixture of all these in more-or-less equal proportions is a loam. The texture of a soil in which one size class predominates is described as, for example, 'clayey' or 'sandy', while compound terms such as 'sandy clay' describe types in which two size classes predominate.

Seasonal or permanent flooding or waterlogging should be noted. Saline or alkaline soils often have a whitish surface crust, and also carry very distinctive vegetation or, locally, none at all.

Soil parent material (the rock or other material from which a soil has developed) may be significant to the plant, but is often difficult to determine, especially in the tropics where many metres of altered material may lie between the soil proper and the unweathered parent rock. The parent materials which most often carry distinctive vegetation are limestone (calcium carbonate or magnesium carbonate), gypsum (calcium sulphate), and serpentine (an igneous rock, often greenish and weathering to a rusty colour, which contains high levels of metals such as nickel). Limestones are common; gypsum is generally only found in arid areas. Serpentines, though rare, often carry distinctive vegetation with endemic plant taxa. Related soils containing large quantities of metals such as copper and cobalt are found in a few parts of the world, and often carry very distinctive vegetation and local endemic species.

Young (1976) is a useful source of information on tropical soils.

Topographic features

The altitude should always be recorded, preferably in metres, although if local maps are marked in feet it is better not to risk errors by converting figures in the field.

Topographical position is often of interest in the description of both plant habitat and soil type. Does the collection point lie on a hill-top, middle slope, lower slope or valley bottom? Finally, the aspect of a sloping site can be important, but less so in the tropics than in the temperate zones, where the microclimate of a south-facing slope may be very different from that of one facing north.

Successional status

Most vegetation is not static – it changes with time. Even mature forest contains gaps produced by falling trees; these slowly fill until they are indistinguishable from the surrounding forest. Observations show that any area of bare ground or open water will be colonized by plants and move through a series of stages until it attains an equilibrium, or *climax,* stage. Field notes can usefully state whether an area of vegetation is in a developing (or *seral)* stage. Particularly in forest, some indication of the time that has elapsed since development began (e.g., since the abandonment of a farm) can be a useful adjunct to field notes.

THE HERBARIUM AND THE ECOLOGIST – THE INTERFACE

As a general rule, it is the ecologist who expects to get information from the herbarium, and he/she should therefore be encouraged to collect good material as well as information useful to the herbarium and to taxonomists working with the collected material. Ideally, the ecologist should discuss with the herbarium staff *before* work starts. Virtually all plant ecological studies require the cooperation of a herbarium.

Ecological projects tend to be of two kinds:

- Community-based (*synecological*) studies, e.g. 'Sand dune vegetation on the coast of Somalia', or 'Regeneration of forest on abandoned farmland'.

- Species-based (*autecological*) studies e.g. 'The biology of savanna mahogany (*Khaya senegalensis*)', or 'The feeding habits of wild rhesus monkeys'.

Ecologists carrying these out tend to be of two types:

- Locally-based personnel: these generally have plenty of time to work on their projects and can visit their sites many times, so that they can return to collect good material if, for instance, there are no flowers on their first visit.

- Short-term visitors: these tend to have little time to carry out their work. They usually expect material to be named from what they can collect at the time of their visit.

Contact ecological personnel whenever possible and emphasize the service a herbarium can provide. Instruct on ways of making a good herbarium specimen. Advise on ways in which good material can be collected at unfavourable seasons or in difficult circumstances. Stress the importance of making voucher specimens and develop a policy on where the specimens are best preserved, see p. 7, *Herbaria for special research programmes*.

Cooperative projects

The ecologist working in the field may have unusual opportunities, such as making observations on pollinators, agents of seed dispersal, and seed predators.

Again, all these require the cooperation of a herbarium, and the preservation of well-annotated specimens.

REFERENCES

Eiten (1974)
Steenis, van (1957a)
Unesco (1973)
White (1983)
Young (1976)

41. CONSERVATION AND THE HERBARIUM

Conservation can be defined as the wise and controlled use of the earth's natural resources. At first sight one may wonder what role the herbarium can play in conservation.

In reality the herbarium has a major and fundamental role. In any field of botanical research, the herbarium is a valuable tool. However, to be useful it should ideally contain material of every species present in the region covered by the herbarium. Such material should have been collected wherever the species occurs in the region. This is particularly important for conservation purposes, since it provides a pattern of distribution for each species with regard to range and possible rarity. It should be remembered that rarity in the herbarium does not equate with rarity in the wild. For example, large trees, such as *Adansonia digitata* in Africa, are easily recognized and so are rarely collected to make herbarium specimens; therefore, the pattern of distribution based on herbarium specimens does not reflect the true distribution of this species. However, rarity in the herbarium can be used as a starting point by providing a check-list from which rare species can be selected. Additional data from the field can then be added to this check-list.

However, a word of warning is necessary. It must be recognized that the overzealous collecting of material for deposit in the herbarium can itself be a threat to the regional survival of species. Therefore a careful assessment of the population size and rarity of the species is needed before whole plants are removed from the wild. Alternatives, such as vegetative propagation or sample seed collection for growing on to provide herbarium material, must be considered. Sadly, herbaria already provide testimony to the extinction of species from given regions and may even record species which have been totally lost to science.

Plant collectors may seek to use herbaria as sources of information for rare species and Curators should always be aware of this possibility. The herbarium therefore provides a range of conservation information, often the essential starting point for conservation action via the check-list of threatened plants. This is the 'red book' approach, where a directory of threatened plants of a particular country or region is produced.

There is another side to the herbarium as a conservation resource which relates to conservation legislation at local, national and international levels. The need for primary lists of rare and threatened species at any level comes first. However, there is a need to interpret these lists. What are the main threats to the species identified, and how best can protection be given to them in the wild? Advice is often needed by those drawing up legislation at whatever level on whether the creation of reserves is an appropriate protection method, or whether specific laws which limit the demand for material from the wild are needed. One may also ask whether the threats are from over-collection for herbal medicine, or for selling off into trade for any commercial reason. These types of threat need different types of legislation.

There are also the consequences of such laws to consider – e.g. identifying plants for the law enforcers and making suggestions on how to provide for the legitimate production of materials of herbal medicine and trade. In addition there is the matter of providing advice on a reserve, together with its size and management. Is it necessary to undertake a rehabilitation programme? If so, what form should that take: seed propagation, tissue culture, removal of plants to another site, etc. Usually it is the herbarium and its scientific contacts that can provide the best data.

Herbaria should accept this role and actively develop it for their local and state administrators. Herbaria should also ensure that they are in the conservation network and receive newsletters from conservation organizations such as IUCN and WWF.

International trade in plant material is in part covered by existing conservation legislation. **CITES** – Convention on International Trade in Endangered Species – which came into existence in 1975, has been ratified by 111 countries (1991). This legislation regulates and controls the trans-shipment of many groups of plants of conservation concern.

Families like Orchidaceae, Cactaceae, Cycadaceae, Cyatheaceae, Dicksoniaceae, Euphorbiaceae (succulent), Aloaceae (Liliaceae) (*Aloe*), Nepenthaceae, Sarraceniaceae, Stangeriaceae, Apocynaceae (*Pachypodium*) and Zamiaceae, in addition to a range of genera and species are covered.

There are three main appendices:

- *Appendix 1*: species from the wild which may not be traded in – with the exception of cultivated material of such species, which are treated like an Appendix 2 species.
- *Appendix 2*: species which may be traded in, if a permit is issued by the management authority of the exporting country and is recognized by the importing country.
- *Appendix 3*: species put forward by countries to which they are native and which are subject to regulation within that country. They are licensed in the same fashion as Appendix 2 species.

All this might seem far removed from the herbarium, but the Convention states that each country which ratifies it must have a management authority and a scientific authority. It is on the advice of the scientific authority that the management authority may issue or monitor licences and permits. This is where the herbarium is involved. Here a herbarium is often the sole source of scientific information on a country's flora and is therefore appointed as the Scientific Authority. Some countries do not appoint a Scientific Authority for plants. Herbaria should in such cases seek to become the national Scientific Authority.

The CITES Convention also applies to movement of material by herbaria. Herbaria as botanical institutions import and export plant material for scientific purposes. The CITES Convention has special provisions which allow streamlined movement of such material. These provisions allow free movement of non-commercial loans, donations or exchanges between CITES *registered scientific institutions* of herbarium specimens, other preserved dried or embedded museum specimens and live plant material which carries an approved label. This provision only applies to movement between *registered* scientific institutions and *both institutions must be registered.* In the case of live plant material it applies to material accessed and integrated in collections and not recently collected material from the wild. All material must carry the appropriate CITES labels. If a herbarium is not registered, movement of CITES specimens is subject to full legal enforcement – it is therefore important to ensure that an active herbarium is registered.

To become registered for CITES purposes, herbaria should contact their national CITES Management Authority. In addition, national CITES Management Authorities hold lists of all registered institutions.

The full text and appendices to the Convention are too lengthy to include here. Information on how the CITES system works in your country and the exact list should be obtained from your national CITES Management Authority. If you do not know who your CITES authority is, this information can be obtained from the CITES Secretariat. Their address is:

CITES Secretariat, 6 rue du Maupas, Case Postale 7, 1000 Lausanne 9, Switzerland.

CITES REFERENCES

The following refer to the mechanisms of the Convention and how it works on a day-to-day basis.

Favre, D.S. (1989). *International Trade in Endangered Species.* Martinus Nijhoff Publishers. Dordrecht, Boston, London. 415pp. ISBN 0 7923 01145 (U.S.)

Lyster, S. (1985). *International Wildlife Law.* Grotius Publications Ltd. Cambridge. 470pp. ISBN 0 906496 22 5.

Wijnstekers, W. (1990). *The Evolution of CITES.* (Revised edition). CITES Secretariat, Lausanne, Switzerland. 284pp. (Available from CITES Secretariat).

REFERENCES

References listed in **22. Essential herbarium literature** and **27. Collectors, itineraries, maps and gazetteers** are not repeated here; see also **p. 91** for references to selected family monographs and **p. 279** for CITES references.

Abbott, L.A., Bisby, F.A. & Rogers, D.I. (1985). Taxonomic analysis in biology. Columbia Univ. Press.

Aguirre León, E. (1986). Epífitas, in Lot & Chiang: 113–119.

Ainsworth, C.G., Sparrow, F.K. & Sussman, A.S. eds. (1973). The Fungi, an advanced treatise. IVa, Ascomycetes and Fungi Imperfecti; IVb, Basidiomycetes and Lower Fungi. New York & London: Academic Press.

Allard, H.A. (1951). Drying herbarium specimens slowly or rapidly. Castanea 16: 129–134.

Allkin, R. & Bisby, F.A. eds. (1984). Databases in systematics. Systematics Association Special Volume 26. Academic Press, London.

— & Maldonado, S. (1991). Workshop: Application of botanical databases within Latin America. Taxon 40: 527–529.

Anon. (1939). The silverfish and firebrat. Economic leaflet no. 3. British Museum (Natural History).

Anon. (1957). Instructions for collectors no. 10. Plants. British Museum (Natural History).

Arber, A. (1938). Herbals, their origin and evolution, 2nd ed.: 138–143. University Press, Cambridge, Mass.

Archer, W.A. (1945). Collecting data and specimens for study of economic plants. United States Dept. of Agriculture misc. publ. 568.

— (1950). New plastic aid in mounting herbarium specimens. Rhodora 52: 298–299.

Arreguín-Sánchez, M. de la L. (1986). Pteridófitas. In: Lot & Chiang: 83–86.

Balgooy, M.M.J. van (1987). Collecting. In: Vogel, E.F. de, Manual of Herbarium Taxonomy, Theory and Practice. Unesco, Jakarta.

Balick, M.J. (1989). Collecting tropical plant germplasm. In: Campbell & Hammond: 476–481.

— (1989a). Collecting and preparing palm specimens. In: Campbell & Hammond: 482–483.

Beaman, J.H. (1965). The present status and operational aspects of university herbaria. Taxon 14: 127–133.

— & Regalado, J.C. (1989). Development and management of a microcomputer specimen-orientated database for the Flora of Mount Kinabalu. Taxon 38: 27–42. (A description of 'LABELS III').

Beatriz Coutiño, B. (1986). Líquenes. In: Lot. & Chiang: 65–73.

Bentham, G. & Hooker, J.D. (1862–1883). Genera Plantarum, 3 vols., London.

Blacklock, C. & Blacklock, N. (1987). Photographing wildflowers, techniques for the advanced amateur and professional. Airlife Publishing Ltd., Shrewsbury. pp. 63.

Botha, D.J. & Coetzee, J. (1976). A portable dryer for herbarium specimens. J.S. Afr. Bot. 42: 41–44.

Brenan, J.P.M. (1968). The relevance of the national herbaria to modern taxonomic research (in Great Britain). In: V.H. Heywood (ed.), Modern Methods in Plant Taxonomy: 23–32. Botanical Society of British Isles and Linnean Society, Academic Press, London.

Brummitt, R.K. (1992). Vascular plant families and genera. Royal Botanic Gardens, Kew. pp. 804.

— & Powell, E. (1992). Authors of plant names. Royal Botanic Gardens, Kew. pp. 732.

Buck, W.R. & Thiers, B.M. (1989). Review of Bryological studies in the Tropics. In: Campbell & Hammond: 485–486.

Burdet, H.M. (1979). Auxilium ad botanicorum graphicem. Conservatoire et jardin botaniques. Geneva. (first published as a series of papers in Candollea, 30–32 (1975–1977)).

Burkill, H.M. (1985). The useful plants of West Tropical Africa. Vol. 1, families A–D. Royal Botanic Gardens, Kew. pp. 960. Vol. 2 E–I is due for publication in 1993.

Burtt, B.L. & Smith, R.M. (1976). Notes on the collection of *Zingiberaceae.* Fl. Mal. Bull. 29: 2599–2601.

Butterfield, F.J. (1987). The potential long-term effect of gamma radiation on paper. Studies in Conservation 32: 181–191.

Bye, R.A. & Linares, E. (1983). The role of plants found in the Mexican markets and their importance in ethnobotanical studies. J. Ethnobiol. 3: 1–13.

Camp, W.H. (1946). On the use of artificial heat in the preparation of herbarium specimens. Bull. Torr. Bot. Club 73: 235–243.

Campbell, D.G. & Hammond, H.D. eds. (1989). Floristic inventory of tropical countries. New York Botanical Garden.

CCI (Canadian Conservation Institute) (1990). List of publications. N = notes; TB = technical bulletins. 1030 Innes. Ottawa, Canada K1A OC8.

Center for Plant Conservation (1986). Recommendations for the collection and *ex situ* management of germplasm resources from wild plants. Missouri Botanical Garden, USA.

Cervera, A.B. (1986). Guía de fuentes de información para el manejo y administración de herbarios. In: Lot & Chiang: 31–44.

Chapman, V.J. & Chapman, D.J. (1973). The Algae. 2nd ed. Macmillan & Co., London, pp. xiv + 497.

Chaudhuri, R.H.N., Banerjee, D.K. & Guha, A. (1977). Ethnobotanical uses of herbaria. Bull. Bot. Surv. India 19: 256–261.

Cholewa, A.F. & Brown, G.K. (1985). Commentary on Compactorization in Herbaria. Taxon 34: 464–467.

Cifuentes Blanco, J., Villegas Ríos, M. & Pérez-Ramírez, L. (1986). Hongos. In: Lot & Chiang: 55–64.

Clark, S. (1986). Preservation of herbarium specimens: an archive conservator's approach. Taxon 35: 675–682.

— (1988). Preservation of herbarium specimens: an archive conservator's approach. Library Conservation News 19: 4–6. A revised version of the above.

Cook, F.E.M. & Hastings, L. (1991). Proposal for standard use descriptions for TDWG. Unpublished text, available from ECOS, Royal Botanic Gardens, Kew.

Cooper-Driver, G.A. & Balick, M.J. (1979). Effects of field preservation on the flavonoid content of *Jessenia bataua*. Bot. Mus. Leafl. 26(8): 257–265.

Coradin, L. & Giannasi, D.E. (1980). The effects of chemical preservations on plant collections to be used in chemotaxonomic surveys. Taxon 29: 33–40 (1980).

Courtecuisse, R. (1991). 7. Protocol for the collection, description and conservation of agaricoid fungi. Flora of the Guianas Newsletter 8, special workshop issue: 39–44.

Cowan, R.S. (1980). Disinfestation of dried specimens at Kew. Taxon 29: 198.

Cremers, G. & Hoff, M. (1990). Constitution et exploration d'un herbier tropical, L'Herbier du Centre ORSTOM de Cayenne: 10–17. ORSTOM, Cayenne.

Crissafulli, S. (1980). Herbarium insect control with a freezer. Brittonia 32: 224.

Croat, T.B. (1978). Survey of herbarium problems. Taxon 27: 203–218.

— (1985). Collecting and preparing specimens of *Araceae*, in Ann. Miss. Bot. Gard. 72: 252–258.

Cronquist, A. (1968). The relevance of the national herbaria to modern taxonomic research in the United States. In: V.H. Heywood (ed.), Modern Methods in Plant Taxonomy: 15–22. Botanical Society of British Isles and Linnean Society, Academic Press, London.

— (1981). An integrated system of classification of flowering plants. Columbia University, New York.

Dahlgren, R.M.T., Clifford, H.T. & Yeo, P.F. (1985). The families of the monocotyledons: structure evolution and taxonomy. Springer-Verlag, Berlin etc.

Dalby, C. & Dalby, D.H. (1980). Biological illustration. Field Studies 5: 307–321 (separate off-print, reprinted in 1985).

Dalla Torre, C.G. de & Harms, H. (1900–1907). Genera Siphonogamarum. Leipzig.

Davidse, G. (1975). Compactors at the Missouri Botanical Garden, a modern system of herbarium storage. Taxon 24: 139–141.

Davis, P.H. (1961). Hints for hard-pressed collectors. Watsonia 4: 6. (1961). (Good practical advice, biased towards drier regions).

Dawes, C.J. (1981). Marine Botany. John Wiley & Sons, New York, pp. x + 628.

Dawson, E.Y. (1956). How to Know the Seaweeds. W.C. Brown Co., Dubuque, pp. 197.

DePew, J.N. (1991). A library media and archival preservation handbook. ABC-CLIO, Inc., Santa Barbara. pp. 441.

De Vogel, E.F. (1987). Manual of herbarium taxonomy, theory and practice. Unesco, Jakarta. pp. 164.

DeWolf Jr, G.P. (1968). Notes on making an herbarium. Arnoldia 28: 69–111.

Digby, J.J. (1985). The collage handbook: 57–61. Thames & Hudson, London.

Dransfield, J. (1979). A manual of the Rattans of the Malay Peninsula: 261. Forest Dept., West Malaysia.
— (1986). A guide to collecting palms. Ann. Miss. Bot. Gard. 73: 166–176.
Duncan, U.K. (1970). Introduction to British Lichens. T. Buncle & Co. Ltd, Arbroath.
Edwards, S.R., Bell, B.M. & King, M.E. eds. (1980). Pest control in Museums, a status report. The Association of Systematics Collection, USA.
Egenberg, I.M. & Moe, D. (1991). A 'stop-press' announcement. Damage caused by a widely used herbarium mounting technique. Taxon 40: 601–604.
Eiten, G. (1974). An outline of the vegetation of South America. In: Symp. 5th Congr. Intl. Primat. Soc.: 529–545.
Engler, H.G.A. & Prantl, K.A.E. (1887–1915). Die Natürlichen Pflanzenfamilien, ed. 1. Leipzig.
Eusebio, M.A. & Stern, W.T. (1964). Preservation of Herbarium Specimens in the humid tropics. The Philippine Agriculturist 48: 16–20.
Faegri, K. & Iversen, J. (1989). Textbook of Pollen Analysis. 4th Edition by K. Faegri, P.E. Kaland, K. Krzywinski. John Wiley, Chichester, pp. 328.
Florian, M.-L. E. (1986). The freezing process – effects on insects and artefact materials. Leather Conservation News, 3: 1–17.
Fosberg, F.R. (1960). Plant collecting as an anthopological field method. El Palacio 67: 125–139.
— & Sachet, M.-H. (1965). Manual for tropical herbaria. Int. Bur. Pl. Tax. & Nom., Regn. Veget. no. 39. Utrecht.
Franks, J.W. (1965). A guide to herbarium practice. The Museum Association, handbook for museum curators, part E, sect. 3.
Fuller, T.C. & Barbe, G.D. (1981). A microwave-oven method for drying succulent plant specimens. Taxon 30: 867.
Gates, B.N. (1950). An electrical drier for herbarium specimens. Rhodora 52: 129–134.
Gleason, H.A. (1933). Annotations on herbarium sheets. Rhodora 35: 41–43.
— & Smith, A.C. (1930). Methods of preserving and arranging herbarium specimens. J. New York Bot. Gard. 31: 112–125.
González-González, J. & Novelo-Maldonado, E. (1986). Algas. In: Lot & Chiang: 47–54.
Grey-Wilson, C. (1980). Notes on collecting *Impatiens*, Fl. Males. Bull. 33: 3435–3436.
Guillarmod, A.J. (1976). Use of odourless carrier, a petroleum product, in preparing herbarium material. Taxon 25: 219–221.
Hall, A.V. (1988). Pest Control in Herbaria. Taxon 37: 885–907.
Hall, D.W. (1981). Microwave: a method to control herbarium insects. Taxon 30: 818–819.
Hallé, N. (1961). Un séchoir à gaz butane pour la préparation des herbiers. J. Agric. Trop. Bot. Appl. 8: 70–71.
Hawksworth, D.L. (1974). Mycologists Handbook. Commonw. Mycol. Inst., Kew.

— , Sutton, B.C. & Ainsworth, G.C. (1983). Ainsworth & Bisby's Dictionary of the fungi, ed. 7. Commonw. Mycol. Inst., Kew.

Haynes, R.R. (1984). Techniques for collecting aquatic and marsh plants. Ann. Miss. Bot. Gard. 71: 229–231.

Henty, E.E. (1976). Notes on the collection of Ferns. In: Womersley: 71.

Heywood, V.H. ed. (1978). Flowering plants of the world, Oxford University Press.

Hicks, A.J. & P.M. (1978). A selected bibliography of plant collecting and herbarium curation. Taxon 27: 63–99.

Hill, S.R. (1983). Microwave and herbarium specimens: potential dangers. Taxon 32: 614–615.

Hollis, D., Jermy, A.C. & Lincoln, R.J. (1977). Biological collecting for the small expedition. Geog. J. 143: 251–265.

Holmgren, P.K. et al. (1990). Index Herbariorum: Pt. 1, The Herbaria of the World, 8th ed. Regnum Vegetabile vol. 120. New York Bot. Gdn., New York.

Holmgren, N.H. & Angel, B. (1986). Botanical Illustration: preparation for publication. New York Botanical Garden, Bronx.

Holttum, R.E. (1957). Instructions for collecting tree ferns. Fl. Males. Bull. 13: 567.

Hutchinson, J. (1926–1934). The Families of Flowering Plants, 2 vols. Oxford University Press.

Hyland, B.P.M. (1972). A technique for collecting botanical specimens in rainforest. Fl. Males. Bull. 26: 2038–2040.

IBPGR (1983). Practical constraints affecting the collection and exchange of wild species and primitive cultivars. IBPGR Secretariat, Rome.

Jain, S.K. & Rao, R.R. (1977). A handbook of field and herbarium methods. Today & Tomorrow's Printers and Publishers, New Delhi pp. 157.

Jeffrey, C. (1973). Biological Nomenclature. Edward Arnold, London.

— (1982). An introduction to plant taxonomy. Cambridge University Press, 2nd. ed.

Jørgensen, P.M., Lawesson, J.E. & Holm-Nielsen, L.B. (1984). A guide to collecting passion flowers. Ann. Miss. Bot. Gard. 71: 1172–1174.

Jury, S.L. (1991). Some recent computer-based developments in plant taxonomy. Bot. J. Linn. Soc. 106: 121–128.

Kobuski, C.E., Morton, C.V., Ownbey, M. & Tryon, R.M. (1965). Report of the committee for recommendations of desirable procedures in herbarium practice and ethics. Brittonia 10: 93–95. Also reprinted in Fosberg & Sachet (1965): 102–104.

Koch, S.D. (1986). Gramíneas y graminoides. In: Lot & Chiang: 93–101.

Kuhlmann, M. (1947). Como herborizar material arbóreo. Secretaria da Agricultura – Estado de Sâo Paulo, Brazil. pp. 39.

Lawrence, G.H.M. (1951). Taxonomy of vascular plants. The Macmillan Company, New York, (especially: 228–262).

Le Thomas, A. (1989). Collection and preparation of pollen samples. In: Campbell & Hammond: 475.

Leuenberger, B.E. (1982). Microwaves: a modern aid in preparing herbarium specimens of succulents. Cactus and Succulent Journal of Great Britain 44(2): 42–43.

Lewis, W.H., Elvin-Lewis, M., Gnerre, M.C. & Fast, W.D. (1988). Role of systematics when studying medicinal ethnobotany of the tropical Peruvian Jivaro. Symb. Bot. Uppsala 28: 189–196.

Lot, A. (1986). Acuáticas vasculares. In: Lot & Chiang: 87–92.

— & Chiang, F. (1986). Manual de herbario, Administractión y manejo de colecciones técnicas de recolección y preparación de ejemplares botánicos. Mexico, pp. 142.

Lundell, C.L. & Kirkham, R. (1966). A method of applying mystox (laurylpentachlorophenate) to protect mounted herbarium specimens. Wrightia 3: 177–180.

Lüning, K. (1990). Seaweeds: Their Environment, Biogeography and Ecophysiology. John Wiley & Sons, New York, pp. xiii + 527.

Mabberley, D.J. (1987). The plant book. Cambridge University Press. (Reissued 1992).

MacDaniels, L.H. (1930). A portable plant dryer for tropical climates. Am J. Bot. 18: 669–670.

Macrander, A.M. & Haynes, R.R. (1990). SERFIS: a methodology for making multi-herbaria specimen databases a reality. Taxon 39: 433–441.

Madalski, J. (1958). Nowa metoda suszenia róslin do zielnika. A new method of plants' drying for herbarium. Fragm. Flor. Geobot. 3: 69–76. Krakow.

Matthew, K.M. (1981). The Carnatic flora project. Reprinted from materials for a flora of the Tamilnadu Carnatic. Tiruchirapalli.

McClean, A.P.D. & Storey, H.H. (1930). A drying cabinet for the preparation of plant specimens for the herbarium. Bothalia 3: 137–141.

McClure, F.A. (1965). Suggestions on how to collect bamboos. In: Fosberg & Sachet: 120–122 & Womersley (1981).

Meikle, R.D. (1971). The history of the Index Kewensis. Biol. J. Linn. Soc. 3: 295–299.

— (1984). Draft Index of Author Abbreviations (2nd. impr.). HMSO.

Michalski, S. (1992). Temperature and relative humidity: the definition of correct/incorrect values. Canadian Conservation Institute (see CCI above), pp. 13.

Mickel, J.T. (1979). How to know the ferns and fern allies. The Pictured Key Nature Series. W.C. Brown Co. Publishers. Iowa.

Millspaugh, C.F. (1925). Herbarium organization. Field Mus. Nat. Hist., Publ. 229, Mus. Tech. Ser. 1.

Mitchell, A.W. (1982). Reaching the rainforest roof – a handbook on techniques of access and study in the canopy. Leeds Philos. Lit. Soc. & UNEP: pp. 36 (over 12 methods described).

Morawetz, W. (1989). Collection and preparation of karyological samples. In: Campbell & Hammond: 469–470.

Mori, S.A., Silva, L.A.M., Lisboa, G. & Coradin, L. (1989). Manual de manejo do herbário fanerogâmico, ed. 2. CEPLAC, Bahia.

Morin, N.R. & others, eds. (1989). Floristics for the 21st Century. St Louis: Missouri Botanical Garden.

Morris, J.W. & Manders, R. (1981). Information available within the PRECIS data bank of the National Herbarium Pretoria, with examples of uses to which it may be put. Bothalia 13: 473–485.

Nicolson, D.H. (1965). In: Fosberg & Sachet: 123–126 & in Womersley (1981).

Nishida, Florence H. (1989). Review of Mycological Studies in the Neotropics. In: Campbell & Hammond: 494–522 (very extensive Bibliography).

O'Shea, B.J. (1989). A guide to collecting bryophytes in the tropics. British Bryological Society Special Volume No. 3. (available from British Bryological Society, National Museum of Wales, Cardiff CF1 3NP).

Page, C.N. (1979). The herbarium preservation of conifer specimens. Taxon 28: 375–379.

Pake, L.V. (1987). Medicinal ethnobotany of Hmong refugees in Thailand. J. Ethnobiol. 7: 13–26.

Pankhurst, R.J. (1983). Preparation of collection labels with a microcomputer. Curator 26: 293–300.

— (1991). Practical taxonomic computing. Cambridge University Press, pp. 202.

Perry, D.H. (1978). A method of access into the crowns of emergent and canopy trees. Biotropica 10: 155–157.

Philbrick, C.T. (1984). Comments on the use of microwaves as a method of herbarium pest control: possible drawbacks. Taxon 33: 73–76.

Pinniger, D. (1990). Insect pests in museums. Archetype Publications Ltd., Denbigh.

— (1991). New developments in the detection and control of insects which damage museum collections. Biodeterioration Abstracts 5: 125–130.

Pohl, W. (1965). Dissecting equipment and materials for the study of minute plant structures. Rhodora 67: 95–96.

Quero, Hermilo, J. (1986). Palmas. In: Lot & Chiang: 121–131.

Quisumbing, E. (1931). Water glass as a medium for permanently mounting dissections of herbarium material. Torreya 31: 45–47.

Radford, A.E., Dickison, W.C., Massey, J.R. & Bell, C.R. (1974). Vascular plant systematics: Chapter 18, Collecting and field preparation of specimens: 387–398 & 31, The herbarium: 751–774. Harper & Row, New York, Evanston, San Francisco, London.

Raynal-Roques, A. (1980). Les plantes aquatiques, Étude et recolte. In: Durand, J.R. & Lévêque, C. eds. Flore et faune aquatiques de l'Afrique Sahelo-Soudanienne. I.D.T. 44: 69–73. ORSTOM, Paris.

Retief, E. & Nicholas, A. (1988). The cigarette beetle *Lasioderma serricorne* (F.) (Coleoptera: Anobiidae): a serious herbarium pest. Bothalia 18: 97–99.

Reznicek, A.A. & Estabrook, G.F. (1986). Herbarium specimens label and annotation printing program for IBM PC computers. Taxon 35: 640.

Roberts, D.A. (1985). Planning the documentation of museum collections. Museum Documentation Assoc. Cambridge.

Roos, M.C. & Verduyn, G.P. (1989). Collecting and germinating fern spores. In: Campbell & Hammond: 471–473.

Rosentreter, R., DeBolt, A. & Bratt, C.C. (1988). Curation of soil lichens. Evansia 5: 23–25.

Ryan, B.D. & McWhorter, F.P. (1986). Processing lichen colonies growing on soil and moss, with glue to facilitate sectioning. Evansia 3: 14–16.

Saccardo, P.A. (1882–1931). Sylloge fungorum omnium hucusque cognitorum, vols. 1–26, Dresden.

Sánchez Mejorada, H. (1986). Suculentas. In: Lot & Chiang: 103–111.

Sarasan, L. (1981). Why museum computer projects fail. Museum News 59: 40–49.

— & Neuner, A.M. (1983). Museum collections and computers. Assoc. of Systematics Collections, Lawrence, Kansas.

Savile, D.B.O. (1962, repr. with addenda 1973). Collection and care of botanical specimens. Canad. Dept. Agr. Res. Branch., Publ. 1113.

Schnell, R. (1960). Technique d'Herborisation et de conservation des plantes dans les pays tropicaux. J. Agric. Trop. Bot. Appl. 7: 1–48.

Schofield, E.K. & Crisafulli, S. (1980). A safer insecticide for herbarium use. Brittonia 32: 58–62.

Shelter, S.G. (1969). The herbarium: past, present and future. Proc. Biol. Soc. Amer. 82: 687–758.

Skvortsov, A.K. (1977). The herbarium. A manual of herbarium methods and techniques (Gerbarii: Posobie po metodike i tekhnike). Nauka, Moscow.

Smith, C. Earle (1971). Preparing herbarium specimens of vascular plants. Agr. Inf. Bull. 348, U.S. Dept. of Agr.

Smith, G.C.A. (1946). A drying cabinet for the herbarium. J. S. Afr. Bot. 12: 43–45.

Soderstrom, T.R. & Young, S.M. (1983). A guide to collecting bamboos. Ann. Miss. Bot. Gard. 70: 12–136.

Stace, C.A. (1989). Plant taxonomy and biosystematics, 2nd ed. Edward Arnold, Hodder & Stoughton, London, pp. 264.

Stansfield, G. (1989). Physical methods of pest control. J. of Biological Curation 1: 1–4.

Stearn, W.T. (1957). An introduction to the Species Plantarum and cognate botanical works of Carl Linnaeus. Facsmile of 1st ed. of Carl Linnaeus, Species Plantarum, pp. xiv + 176. Ray Society, London.

— (1983). Botanical Latin. Nelson. London, pp. xiv + 566.

Steenis, C.G.G.J. van ed. (1950). The technique of plant collecting and preservation in the tropics. In Steenis-Kruseman, M.J. van, Flora Malesiana 1: XLV–LXIX.

— (1957). The policy and value of distribution duplicates of tropical collections. Taxon 6: 133–135.

— (1957a). Outline of vegetation types in Indonesian and some adjacent regions. Proc. Pac. Sci. Congr. 4: 61–97.

— (1962). Interaction and cooperation between tropical and temperate herbaria. Proc. 9th Pac. Sci. Cong. 1957. 3: 54–55.

— (1972). Reaping the harvest of names and identification by means of identification and collection lists. Fl. Males. Bull. 26: 2020–2037.

— (1977). Three pleas to collectors – improve your field data. Fl. Males. Bull. 30: 2843–4.

Stern, W.T. (1965). Wood collections. In Fosberg & Sachet: 63–64.

Stolze, R.G. (1973). Inadequacies in herbarium specimens of large ferns. Amer. Fern J. 63: 25–27.

Stone, B.C. (1983). A guide to collecting *Pandanaceae* (*Pandanus*, *Freycinetia* and *Sararanga*). Ann. Miss. Bot. Gard. 70: 137–145.

Strang, T. (1992). Museum pest management. Canadian Conservation Institute (see CCI above), pp. 26.

Subramanyam, K. & Sreemadhavan, C.P. (1970). The paramount role of herbaria in modern taxonomic research. Bull. Bot. Surv. India 12: 210–212.

Taylor, P. (1977). On the collection and preparation of *Utricularia* specimens. Fl. Males. Bull. 30: 2831–2832.

— (1988). The genus *Utricularia*, a taxonomic monograph. Kew Bulletin Additional Series XIV: 54–55.

ter Welle, B.J.H. (1989). Collection and preparation of bark and wood samples. In: Campbell & Hammond: 467–468.

Tétreault, J. & Williams, R.S. (1992). Materials for exhibit, storage and packing. Canadian Conservation Institute (see CCI above), version 4.1, pp. 13.

Tomlinson, P.B. (1965). Collecting botanical material in fluid preservatives, with special reference to the tropics: 117–119 & Special techniques for collecting palms for taxonomic study: 112–116. In: Fosberg F.R. & Sachet.

Touw, M. & Kores, P. (1984). Compactorization in Herbaria. Taxon 33: 276–287.

Turrill, W.B. (1964). Plant taxonomy, phytogeography and plant ecology. In: Turrill (ed.), Vistas in Botany 4: 187–224.

Unesco (1973). International classification and mapping of vegetation. Unesco, Paris. [Parallel texts in English, French and Spanish].

Veldcamp, J.F. (1987). Botanical nomenclature of vascular plants. In: De Vogel: 127–146.

Wendt, T. (1986). Árboles. In: Lot. & Chiang: 133–141.

West, K. (1983). How to draw plants. The techniques of botanical illustration. The Herbert Press in association with The British Museum (Natural History). London.

Wetmore, C.M. (1979). Herbarium computerization at the University of Minnesota. Syst. Bot. 4: 339–359.

White, F. (1983). The vegetation of Africa, a descriptive memoir to accompany the Unesco/AETFAT/UNSO vegetation map of Africa. Unesco, Paris.

Whitmore, T.C. & Fosberg, F.R. (1965). Lauryl pentachlorphenate protecting herbarium specimens. Taxon 14: 164–166.

— (1972). The description of a tree. Chapter 2 *in* Tree Flora of Malaya, Vol. 1. Longman Group Ltd., London.

Williams, D.W. (1987). The guide to museum computing. AASLH (Amer. Assoc. for State and Local History, Nashville).

Willis, J.C. (1973). A dictionary of the Flowering Plants and Ferns, 8th ed., by H.K. Airy Shaw. Cambridge Univ. Press.

Womersley, J.S. (1957). Paraformaldehyde as a source of formaldehyde for use in botanical collecting. Rhodora 59: 299–303.

— (1976). Plant collecting for Anthropologists, Geographers and Ecologists. Papua New Guinea Bot. Bull. 2, ed. 2. Original ed. 1969.

— (1981). Plant collecting and herbarium development. FAO Plant production and protection paper 33. Rome. pp. xi + 137.

Wood, P. (1979). Scientific illustration. Van Nostrand Reinhold. New York, Cincinnati, Toronto, London, Melbourne.

Young, A. (1976). Tropical Soils and Soil Survey. Cambridge Univ. Press.

Zweifel, F.W. (1961). A handbook of biological illustration. University of Chicago Press, Chicago.

Zycherman, L.A. & Schrock, J.R. (1988). A guide to museum pest control. Foundation of the American Institute for conservation of Historic and Artistic Works and the Association of Systematics Collections, Washington.

INDEX